WARNINGS

W A R N I N G S

THE TRUE STORY OF HOW SCIENCE
TAMED THE WEATHER

MIKE SMITH

GREENLEAF
BOOK GROUP PRESS

Published by Greenleaf Book Group Press
Austin, Texas
www.gbgpress.com

Copyright ©2010 Mike Smith Enterprises, LLC
www.mikesmithenterprises.com
All rights reserved.

Distributed by Greenleaf Book Group LLC

For ordering information or special discounts for bulk purchases, please contact Greenleaf Book Group LLC at PO Box 91869, Austin, TX 78709, 512.891.6100.

Design and composition by Greenleaf Book Group LLC
Cover design by Greenleaf Book Group LLC

Publisher's Cataloging-In-Publication Data
(Prepared by The Donohue Group, Inc.)

Smith, Mike (Michael Ray), 1952-
    Warnings : the true story of how science tamed the weather / Mike Smith. -- 1st ed.

    p. : ill. ; cm.

    ISBN: 978-1-60832-034-9

1. Weather forecasting--United States--History. 2. Meteorology--United States--History. 3. Weather forecasting--United States--Technological innovations. 4. Meteorological services--United States--Technological innovations. 5. Tornadoes--United States. 6. Smith, Mike--(Michael Ray), 1952- 7. Meteorologists--Biography. I. Title.

QC995 .S65 2010
551.6/3   2010920227

Part of the Tree Neutral™ program, which offsets the number of trees consumed in the production and printing of this book by taking proactive steps, such as planting trees in direct proportion to the number of trees used: www.treeneutral.com

Printed in the United States of America on acid-free paper

10 11 12 13 14   10 9 8 7 6 5 4 3 2 1

First Edition

*On a personal note, I would like to dedicate this book to my family:
my amazing wife, Kathleen, and my wonderful children,
Richard; Brandon and his wife, Mary; and Tiffany.
Thank you for your love and encouragement!*

*Professionally, I dedicate this book to my talented colleagues at
WeatherData® and AccuWeather® and to the larger community
of applied meteorologists in the private sector, the National
Weather Service, and broadcast meteorology, along with storm
spotters and emergency managers, who do not get nearly enough
credit for their extraordinary dedication and the tens of
thousands of lives they have saved.*

CONTENTS

ACKNOWLEDGMENTS

A SPECIAL THANK YOU to Michelle Strecker and Kim Dugger. Many people and organizations, directly or indirectly, contributed to this work. Most are named within the manuscript. Others include Jeanette Cezanne, Rick Durden, Stan Vannier (Union Pacific Railroad), Steve Pryor, Jim Reed, Kaz Fujita, Katherine Bay, Weatherwise, Don Burgess, Marlene Bradford, Dennis Feltgen, Thomas Murphy, Larry Schwam, Dick McGowan, Matt Dennis, Karen Ryno, the U.S. Air Force, www.tornadochaser.com/UDALL, Udall Historical Museum, texasbar.com, stormscapesdarwin.com, "Caught in the Path," the National Climatic Data Center, FEMA, and NOAA (especially the Wichita NWS and NSSL). If I have left someone out, I apologize! I also want to thank all of the people I interviewed as I wrote the manuscript. Some did not want their names used, which is why I am not listing them individually.

 When you see this symbol, go to www.warningsbook.com for videos, related info, and more.

PHANTOM ACCIDENTS

PICTURE THIS.

On October 9, 2005, passengers boarded Amtrak's *Vermonter*, a once-a-day train that runs between Washington, D.C., and St. Albans, Vermont, a small rustic town just a few miles from the Canadian border. Though the Vermont weather had been stormy for several days, the passengers expected a routine, relaxing ride on this Sunday morning.

But the passengers were grossly mistaken: this trip was to be anything but routine.

Along its route in the Green Mountain State, the *Vermonter* rides the rails of the New England Central Railroad at a brisk 59 miles per hour. Near the town of Putney, the train comes around a curve and dips slightly as it descends a short grade into a narrow river valley.

On this morning, as the train rounded the curve and the river came into view, the locomotive's engineer and conductor were horrified to see a washout ahead. About one hundred feet of iron rails and cross ties were hanging in the air with absolutely nothing supporting it; a flash flood had washed away the track's support structure.

The engineer instantly applied the emergency brakes. Metal wheels ground against metal rails. Sparks flew as the brakes squealed. In the locomotive's cab, desperate prayers were offered that the train would stop in time.

One hundred feet left to the washout, then fifty . . . twenty . . . ten. The momentum of the passenger cars and locomotive made it difficult for the train's wheels to gain traction. With the downward slope of the track, gravity was working against the brakes. Standing passengers were thrown forward in the rapid deceleration. Sitting passengers tried to grab the seats in front of them so they wouldn't fly headfirst. Hot coffee became airborne.

The train was, in railroad-industry jargon, "in emergency." A radio signal from the locomotive, triggered by the emergency brake application, caused a bright light to flash in the dispatch center in St. Albans. The dispatcher called the train to learn the cause of the emergency. The crew didn't have time to answer.

For an instant, as the train's wheels ventured over the unsupported track, both locomotive crew members thought the rails might hold. But when the locomotive had coasted nearly thirty feet over the washout, one of the steel rails snapped, plunging the train into the water. The engineer and conductor were killed instantly as the locomotive landed nose-first. One passenger car, then a second, fell in. A third car was left dangling perilously from the track.

When the wreckage had been cleared, forty-seven people were dead. Family after family made the dreaded call to priests, rabbis, and ministers to plan funerals. Some survivors were crippled for life. Attorneys began preparing lawsuits. The tragedy was national news for days.

* * *

Here's the good news about this story: it never happened.

A seemingly unavoidable tragedy (once the train crew saw the washout, they never could have stopped the train in time) was avoided. The reason why this scenario was a phantom—rather than a real—accident is the subject of this book.

Some things about this story are true. There is a *Vermonter*. It travels at a speed of 59 miles per hour. There was heavy rain on October 9, 2005. That day the track's supports washed out with nearly one hundred feet of rail left dangling in the air. And the train was operating that day—for a time.

The *Vermonter* didn't have a heavenly guardian angel on that October day. What it did have, then and every day since then, is an earthly one: people sitting in a room, much like air-traffic controllers, staring at sophisticated computer displays and making sure that trains (which may need more than a mile of track to stop) have sufficient warning to avoid disasters such as the imagined one here.

The Vermont Washout. Courtesy of the New England Central Railroad.

Those sentries work for WeatherData, a company of meteorologists and other scientific professionals. In this case, they provided

flash-flood warnings to the American Rail Dispatch Corporation
(ARDC), which relayed the warnings to the New England Central
Railroad. The railroad sent out crews to inspect the track ahead of
the train and found the washout. The train was stopped in time.

Would forty-seven people have died if the *Vermonter* had reached
the washout? The answer is likely yes. That was the death toll when
an Amtrak train, the *Sunset Limited*, plunged off a bridge in Alabama
in 1993 when it encountered a misaligned track in dense fog.

What really happened that day in Vermont is much less dramatic,
but far more important. The *Vermonter* was canceled. Grumbling
and complaining passengers were put on buses to their destinations.
These redirected, disgruntled passengers perhaps never fully com-
prehended the tragedy that had been averted. Tom Murphy of ARDC
told the *Wall Street Journal* a few days later that if the train had gone
over the washout, "it would have been a catastrophe."

* * *

How many times have you seen something like this on your television
screen: "From Eyewitness News . . . A flash-flood warning is in effect
for Washington County until 11:00 p.m. . . . ."

That advisory, known as a "weather crawl," is a part of everyday
life for most people. As I write these words, my local CBS television
affiliate is running a crawl over its telecast of the AT&T PGA golf
tournament in Pebble Beach about an approaching winter storm with
torrential rain and snow. Weather information surrounds us to such
an extent that it is almost completely taken for granted. So much so
that viewers of TV stations sometimes call to complain that the crawl
diminishes their enjoyment of the program they are watching.

While it's probably human nature to complain about all types of
inconveniences, major, minor, and imagined, I suspect the people
who call to complain about storm warnings have never really given

the subject much thought. It's just weather, right? They probably assume flash floods don't pose much of a threat. So why interrupt the TV program?

Actually, flash floods—to take just one type of storm as an example—can (and do) kill thousands of people. In Johnstown, Pennsylvania, 2,209 people died in a single flash flood in 1889. (By comparison, 2,280 people died in the attack on Pearl Harbor.) While fatality numbers of previous flash-flood disasters, such as the Johnstown flood, are significant, given today's larger and often denser urban populations, the hydrologic threat from flash floods is greater than ever.

And it's not just flash floods we should fear. In 1900, before a comprehensive warning system existed, the Galveston hurricane caused the deaths of about 8,000 people—19 percent of the population of the island, and almost *three times* the number killed in the 9/11 terrorist attacks. But consider this contrast: when Hurricane Andrew (*stronger* than the Galveston hurricane) struck modern-day south Florida in 1992, the death toll was only twenty-four. Those deaths represented only a tiny percentage of the 4.2 million people affected by the storm.

Before the current tornado warning system was in place, tornadoes causing triple-digit fatalities were surprisingly common, with the worst of them killing 689 people.

Perhaps those crawls on TV serve a purpose, after all.

* * *

To really understand the success of meteorological forecasting, it helps to put it in context. We all know that science is constantly working to save lives, right? So, pop quiz: Which area of science do you think has been the most effective at saving lives over the past fifty years?

Well, let's take a look:

1. Cancer research. On December 23, 1971, President Nixon declared a "war on cancer." More than $300 billion later (allowing for inflation), cancer rates (deaths per 100,000 of U.S. population) are *up*.

2. Heart disease research. Heart disease death rates were down 50 percent from 1980 to 2000, according to a study in the *New England Journal of Medicine*. While this news is both welcome and impressive, it has been purchased at an extremely high cost. According to the Centers for Disease Control, spending on cardiac care and research was $394 billion in 2005, the most recent year for which figures are available.

3. Traffic safety. Air bags, seat belts, child safety seats, specially engineered steering columns, safety glass, shoulder harnesses, dual braking systems, and so forth were not standard equipment on 1950s cars. Since the creation of the National Traffic Safety Administration in 1966, deaths per 100,000 of U.S. population are down about 40 percent. This is clearly a positive development, but the costs (tax dollars, increased cost of automobiles, research funding) have been in the hundreds of billions of dollars.

So how does meteorology, the science of weather, stack up?

The truth is that many people don't even think of meteorology as a science. Some college physics professors have been heard to denigrate their meteorology colleagues, alleging that meteorology is not a "real" science. After all, there is a Nobel Prize for physics. There is a Nobel Prize for chemistry. There is a Nobel Prize for medicine.

There is no Nobel Prize for meteorology.

Yet if we go back to our pop quiz, it seems that meteorology is the science that's most successful at saving lives. In the 1920s, the annual death rate from tornadoes in the U.S. was approximately three per million people. In the early 1950s, with the beginning of a tornado warning system, the rate was still 1.5 deaths per million people. In the last three years, 2006 through 2009, the death rate was down to .068 deaths per million, a decrease of more than 95%! We've never needed a presidential "war on tornadoes," simply because meteorologists have quietly taken care of it. They've brought us from seeing tornado deaths as an "inevitable part of population increase" to where we are today, with the lowest tornado-related death rates ever.

What we see, in fact, is a far more impressive performance in reduction of deaths by meteorology than that offered by cancer research, heart disease research, or traffic safety initiatives. The budgets for storm research have been a tiny fraction of those allocated to these other areas, so this increased safety has come at a very low cost to American society.

Similarly impressive statistics showing a decline in deaths could be assembled for hurricanes, floods, and other storms. Commercial aviation is the only field that comes close to meteorology in the rate of reduction of fatalities, but even this progress is partly due to weather science.

Here are some examples. From 1964 to 1985, a number of microburst-related commercial plane crashes in the United States killed hundreds of people at a time. (A microburst is a form of extreme wind shifts nearly impossible to safely traverse.) Today, this type of fatal airline accident has practically been eliminated.

From 1986 to 2008, the number of microburst fatalities in the United States was 37, a decrease of 93 percent, in spite of a near doubling of airline flights during this period. The last fatal commercial aviation crash involving a microburst was in July 1994.

Every time a jumbo jet breaks off a landing approach after receiving a "wind shear" alert, a near certain crash turns into a phantom, with more than one hundred lives saved.

Innovations in meteorology over the past fifty years are saving countless lives on an almost weekly basis and make the American economy run more smoothly and profitably.

So why isn't this common knowledge? Because catastrophes that
don't happen aren't noticed let alone considered newsworthy. It took
Sherlock Holmes's considerable powers of deduction to realize the
importance of a dog *not* barking. It would take the great detective's
powers of observation to similarly note the planes that don't crash,
the trains that aren't derailed, the people who aren't killed.

Yet saving lives and property is routine for meteorologists.

Popular culture and entertainment never make a hero of the
meteorologist; if anything, he is treated as a buffoon, the weatherman who can never get it right. The reality is that we *are* getting it
right, and getting it right when it counts the most.

The only thing standing between the American public and huge
annual death tolls from extreme weather is the weather warning system—a partnership of weather companies, the federal government,
emergency managers, volunteer storm spotters, and the media. As
coastal populations grow in areas vulnerable to hurricanes, as more
people move into flood plains, and as megacities grow in tornado-
prone areas, the warning system becomes more and more important.

* * *

This weather warning system didn't always exist. In fact, for years the
U.S. government had strict rules *against* forecasting tornadoes and
hurricanes.

In 1895, Willis Moore became chief of the Weather Bureau. He
brought with him an obsessive fear of his agency misforecasting
major storms. His solution? Don't forecast any! He decreed that
the word hurricane was not to be included in a forecast, unless it
was used to *deny* a hurricane existed. Here's an example: "A storm,
not a hurricane, will cause rainy conditions over Florida the next few
days." (This practice is chronicled in Erik Larson's excellent book
Isaac's Storm.)

Willis Moore also closed down a tornado forecasting program that had begun during the previous decade and which had shown some promise. And he banned the use of the word *tornado* in Weather Bureau forecasts. This ban was reinforced by the Bureau several times in the following decades and was still in effect in the 1950s.

But the tornadoes didn't care. Young Elvis Presley nearly died in a 1936 tornado in Tupelo, Mississippi, that killed well over 200 people. In 1953 alone, individual tornadoes killed 94 people in Massachusetts, 115 in Michigan, and 114 in Texas.

Rather than surrender to these ever-increasing death tolls, a courageous meteorologist successfully defied the ban on tornado warnings, ushering in a major change in attitude toward meteorology in the late 1950s. Critical research was begun, often with war surplus–salvaged parts and on shoestring budgets, which would result in reshaping our weather-forecasting culture.

And I've been privileged to have a front-row seat.

Applied meteorology is a small field with only about ten thousand practitioners. I am honored to be one of them, and I have met many of the remarkable people whose work has created the new science of storm warnings. This is the story I will tell in the pages that follow: where we began, how far we've traveled, and where we still need to go.

Make sure you fasten your seat belt, though: it's a turbulent ride.

CHAPTER ONE

THE RUSKIN HEIGHTS TORNADO

MY FATHER BURST THROUGH THE FRONT DOOR. *"Here it comes!"*

Mom rushed down the hallway of our modest ranch house to grab my baby brother Phillip out of his crib so we—she, my dad, my other brother Mark, and I—could scamper down the stairs into our basement.

Dad had been standing in the front yard with a neighbor, watching a strange green sky. I'd seen that sky, too: my face had been plastered up against our large picture window until Mom had pulled me back. "They say to keep away from windows!" she had scolded.

But I was five years old and didn't want to keep away. The clouds were like nothing I'd ever seen.

During the hour or so before that frantic dash down the stairs, the local TV station routinely interrupted *I Love Lucy* with an announcer telling us to turn off the gas and head to the southwest corner of the basement. Soon enough, the five of us were huddled in exactly that location.

That was when I started to get scared.

* * *

Let's backtrack a little.

May 20, 1957, started out hazy in Kansas City. The late-May lawns were green, and the trees were leafy. On the suburban south side of town, just about every backyard had a clothesline. Monday was laundry day throughout the neighborhood, and at house after house, shirts, pants, and underwear were all hanging on the line to dry. When Mom hung out her freshly laundered bed sheets, they flapped in the wind and strained against the clothespins that fastened them to the line.

On that morning, laundry dried slowly because there was a dull overcast sky and winds were light from the east. Temperatures were in the sixties. Throughout Kansas City, people went to work, planned graduation parties, and listened to baseball on a newfangled device, the portable radio.

Around noon there was a sudden change in the weather. The winds shifted to the south and gusted to 35 miles per hour. Temperatures and humidity rose rapidly. By mid-afternoon the air had an ominous, heavy feel to it. Longtime residents knew that when the air had felt that way in the past, bad things were usually on their way.

School let out at three fifteen. As car radios played "Canadian Sunset" and "My Prayer," mothers drove their children home, perhaps stopping first at the grocery store. If they were listening when the four o'clock news broadcast came on, they learned there had been tornadoes in central Kansas earlier in the day. Given the muggy feel of the air, some of them silently resolved to keep a radio or television on in case severe thunderstorms developed in Kansas City.

Once the kids were home, they did homework, played outside, or watched *Adventures of Superman* or *The Mickey Mouse Club* on TV. No one knew, yet, what was coming.

* * *

Eleven miles to the north, in a small tower on the second floor of the Kansas City Municipal Airport, Weather Bureau meteorologists were changing shifts.

The day shift, including the Bureau's best radar operator, Joe Crites, had been closely monitoring the storms to the west and northwest of Kansas City. Now coming on duty for the four-to-midnight shift were Bob Babb and Joe F. Audsley. It was Babb's job to make the local forecast and Audsley's to watch the radar and record "observations"—airport weather conditions—for pilots and the media. Because Babb's official title was public service forecaster, he handled the phones.

During a major weather event, this was a lot of work for two people. Due to the possibility of severe storms developing in the area, Crites volunteered to stay and assist Babb and Audsley, but his staying late would mean overtime—something frowned on by management. Crites then offered to stay and work without pay, but management turned him down a second time. After all, this was an agency so conservative it was widely rumored that each year it returned at least part of its appropriations to the U.S. Treasury.

Meanwhile, across the city breadwinners picked up the evening paper on their way home from work. With television still in its infancy, most people depended on the afternoon newspaper to learn of important events. On this day, however, the paper wouldn't tell them what they needed to know about the weather changes that were coming.

At the Weather Bureau, the converted World War II radar depicted showers immediately west of Kansas City and a strong thunderstorm near Leavenworth. Other thunderstorms were scattered throughout northeast Kansas and northwest Missouri.

Joe Crites was still hanging around at 4:08 p.m. when a small, white smudge first appeared near the southwesternmost edge of the radar display. At first the forecasters weren't sure whether they were seeing anything. It took a couple more 360-degree sweeps of the antenna before they were confident that something was definitely appearing on the screen. Audsley and Crites looked at each other. Crites pointed to the tiny smudge on the screen indicating a thunderstorm in an area about fifteen miles northeast of Wichita and said, "If Kansas City is going to have a problem, it will be *that* one." Reluctantly, he walked out the door.

Joe Audsley and Bob Babb were on their own.

In 1957, figuring out precisely where a storm was located was a challenge. Unlike the colorful and highly detailed radar images we see on television today, the Bureau's radar screen was a simple display of white blotches with bright narrow circles (called "range rings") designating 50-mile intervals. The Kansas City Weather Bureau used standard road maps from gasoline stations to approximate the radar blotches to the storms' physical locations.

Two minutes after the radar echo first appeared northeast of Wichita, the Weather Bureau's national severe weather forecasting center, also located in Kansas City in a high-rise federal office building just nine blocks from the airport weather station, issued a "severe weather forecast" that included the Kansas City area.

From the national severe weather forecasting center:

SEVERE WEATHER FORECAST ISSUED BY THE WEATHER BUREAU FORECAST CENTER KANSAS CITY MO . . . 4:10PM CST MONDAY MAY 20 1957 . . . SEVERE WEATHER FORECAST NUMBER 169 . . . SCATTERED SEVERE THUNDERSTORMS AND SEVERAL TORNADOES FOR THE REST OF THE AFTERNOON AND UNTIL 9PM THIS EVENING.

The little radar echo northeast of Wichita was growing in size and brightness and was moving rapidly toward Kansas City.

* * *

On this day, some south suburban Kansas City routines would vary from the typical Monday evening's schedule. It was kindergarten graduation in the town of Martin City, and the annual sports banquet at St. Catherine's School was being held in nearby Hickman Mills. There was a graduation rehearsal at Ruskin Heights High School.

Ruskin was a rapidly growing suburb modeled after New York's Levittown. The builders Praver & Sons erected prefabricated homes marketed primarily to World War II and Korean War veterans. A three-bedroom, one-bath ranch house built on a flat concrete slab cost $11,200. A basement was an additional $1,500, an expense that few new homeowners incurred. In an area that had literally been cow pasture four years earlier, Praver poured as many as twenty-five slabs a day, which added up to almost 1,900 homes in 1957.

The Weather Bureau—including the local and national offices— was connected to Kansas City media with a teletype system informally known as the "local loop." At 5:30 p.m., Joe Audsley sent the following notice, which printed at a speed of about fifteen words per minute.

From the local Weather Bureau office at the Municipal Airport:

```
AT 5:30PM . . . A SEVERE THUNDERSTORM AT
EMPORIA KANSAS WAS GIVING HAIL UP TO ONE
INCH IN DIAMETER. RADAR AT THE KANSAS
CITY WEATHER BUREAU SHOWS THIS STORM
TO BE VERY SEVERE AND MOVING NORTHEAST-
WARD IN THE GENERAL DIRECTION OF KANSAS
CITY AT ABOUT 50MPH . . . AT THE PRESENT IT
APPEARS THAT KANSAS CITY MAY EXPECT HIGH
WINDS ACCOMPANIED BY HAIL BY 8:00PM OR
SHORTLY BEFORE. END. JFA
```

Where the number of homes rapidly grows, commerce soon follows. The brand-new Ruskin Heights Shopping Center included an A&P grocery store, a hardware store, a branch of the Jackson County

Public Library, and several other shops. Northeast of the shopping center was the Ruskin High School, also brand new. The following evening, it planned to graduate its first students.

From the local Weather Bureau office at the Municipal Airport:

```
AT 6:05PM . . . THE SEVERE THUNDERSTORM
MENTIONED AT 5:30PM AT EMPORIA KANSAS
IS NOW CENTERED 55 MILES SOUTHWEST OF
KANSAS CITY AND CONTINUES TO MOVE IN THIS
DIRECTION. IT IS . . . STILL SEVERE. FURTHER
ADVICES WILL FOLLOW. JFA
```

Just after six o'clock, Audsley had observed a hook echo form on the radar—the fishhook shape that told him a tornado was in progress—and he composed the above message. Ironically, he couldn't mention the word tornado—that word was reserved for use by the national severe weather center. Field offices, like the one at the Kansas City Municipal Airport, were not allowed to use it.

At 6:15 p.m., a tornado touched down just outside the small town of Williamsburg, about sixty miles southwest of Kansas City, and began moving northeast. Its first destruction occurred at the U-Rest Hotel and Restaurant near Ottawa. Fortunately, the hotel owner saw the tornado coming and evacuated the patrons in his family's station wagon.

Back in Kansas City, Joe Audsley recorded the latest weather information. Conditions downtown were still unremarkable.

From the local Weather Bureau office at the Municipal Airport:

```
6:25PM . . . TEMPERATURE 80. HUMIDITY 60.
WIND SOUTH 20. CLOUDY LIGHTNING WEST. SEA
LEVEL PRESSURE 29.43 UNSTEADY. MAXIMUM
TEMP TODAY 83. MINIMUM 52. NO PRECIPITA-
TION. JFA
```

Meanwhile, at my house we were finishing dinner as the sky began to darken.

News anchor Randall Jesse of WDAF Radio and TV had gone home for the day but was called back to the TV station where he found the newsroom in organized chaos, with police radios crackling, telephones ringing, and the local loop weather wire clattering out bulletins.

Ruskin Heights tornado passing near Ottawa, Kansas, May 20, 1957. Courtesy *The Ottawa Herald.*

From the local Weather Bureau office at the Municipal Airport:

AT 6:30PM A FUNNEL CLOUD TOUCHING THE GROUND WAS SIGHTED 40 MILES SOUTHWEST OF KANSAS CITY MOVING NORTHEASTWARD. AT 6:52PM ANOTHER FUNNEL CLOUD WAS SIGHTED IN THE VICINITY OF PAOLA KANSAS MOVING RAPIDLY NORTHWARD . . . THE FORECAST FOR KANSAS CITY CALLS FOR THUNDERSTORMS, POS-

SIBLY SEVERE, WITH A CHANCE OF A TORNADO
IN THE GREATER KANSAS CITY AREA.

A "funnel cloud touching the ground" is, in fact, a tornado, and as sightings poured in, the local office started using the word tentatively, not to issue a "warning" but to reinforce the national office's *forecast*.

From the local Weather Bureau office at the Municipal Airport:

TORNADO REPORTED AT SPRING HILL ABOUT 20
MILES SSW OF KANSAS CITY.

The tornado struck the home of the Davis family of Spring Hill. All four family members were killed. It continued northeast, mainly in open country, as it approached the Missouri border. None of the previous Bureau messages contained the word *warning*, and none of them urged people to take shelter ahead of the storm. The institutional fear of setting off a panic was so pervasive in the Weather Bureau culture of 1957 that a simple message, such as "A tornado is coming, take shelter!" was unthinkable.

About seven o'clock, my family was settling in to watch *I Love Lucy*, just as we did every Monday night. But the show was repeatedly interrupted with weather bulletins. At some point we started flipping between the channel televising *I Love Lucy* and channel four, WDAF, owned by the *Kansas City Star* and widely favored for local news.

Randall Jesse, Walt Bodine, and other news anchors were now on the air almost continuously reading weather bulletins.

Back at the Weather Bureau, Audsley was tense; like other experienced meteorologists, he could sense when something terrible was going to happen. Audsley had been in dangerous situations before; he'd been the Navy's weatherman at the World War II battles of Iwo Jima and Okinawa. Even those ordeals did not prepare him for the pressure of what was to come. He later described himself as having a pain in his stomach as he watched what had been a dim, white smudge near Wichita three hours earlier start closing in on Kansas City.

What we now call a tornado watch had been issued for Kansas City at 4:10 p.m. by the national severe weather center, but there were no official procedures to follow if Audsley and Babb believed a tornado to be *imminent*. The hook echo, which had been indistinct for a while, returned to the radar screen at 7:23 p.m., just miles southwest of Kansas City.

Ten years earlier, Audsley had issued a message about a potential tornado when he was stationed in Sioux City, Iowa. He was called into his supervisor's office the next day and reprimanded, instructed to keep quiet about the incident so that "maybe the region [Weather Bureau Central Region Headquarters, to which the Sioux City office reported] will never hear about it." The fact that Audsley's alert turned out to be correct did not matter: the Weather Bureau did not want to be in the tornado warning business.

Audsley had to be remembering that reprimand when he overheard Bob Babb take a call from the Naval Air Station on the south side of Olathe, Kansas, alerting them of a tornado on the ground moving rapidly northeast. Audsley typed,

```
AT 7:23PM RADAR AT THE AIRPORT SHOWS AN
ECHO WHICH APPEARS TO BE VERY SEVERE
JUST 3 OR 4 MILES SOUTHEAST OF OLATHE
KANSAS MOVING NORTHEASTWARD. WE HAVE
JUST THIS MOMENT RECEIVED A REPORT OF A
TORNADO ON THE GROUND MOVING RAPIDLY
NORTHEASTWARD AT THIS EXACT SPOT. END.
JFA
```

Immediately after sending the message, Audsley and Babb turned to each other and said, nearly in unison, "What have we done?" They'd used the forbidden word, and by reporting that it was moving "rapidly northeast" had done the closest thing to issuing a message to take cover—what we now call a tornado warning.

Even though the sun had not yet set, a foreboding green overcast spread darkness over the south part of the city. As the bulletin was received in newsrooms, TV and radio stations reported Audsley's warning in particularly urgent tones.

It sent people throughout the south Kansas City area scurrying. Some went outside to try to see the tornado, others to their basements, and still others to *find* a basement. Now that the tornado was in the metropolitan area, newsroom police and fire scanners crackled to life with sightings of the tornado and reports of debris flying through the air. The Kansas City TV stations were now in continuous weather coverage, probably the first time in broadcast history this had ever occurred.

From the national severe weather forecasting center:

```
AIRLINE PILOT REPORTS FUNNEL CLOUD TWO
MILES WEST OF GRANDVIEW AIR FORCE BASE
AT 7:37PM.
```

As the Frontier Airlines pilot watched, the tornado crossed the line separating Kansas from Missouri. The tornado had traveled through mostly rural areas in Kansas, killing seven people along the way; but now that it was in Missouri, the number of people in the tornado's path was in the tens of thousands.

First up was Martin City.

A stop on the Missouri Pacific Railroad, Martin City considered itself a separate town, rather than a Kansas City suburb. Railroad tracks cut diagonally through the town just west of the main intersection. As darkness deepened, a distant but growing rumble was heard, dismissed by many hearing it as just another train. But when sudden lightning turned the stormy darkness into light, the dreaded funnel cloud was revealed.

If a tornado appears to not be moving but *does* appear to be growing larger, then it is coming right at you. With each lightning bolt, the funnel cloud appeared larger than it had during the flash before.

A block northwest of the Martin City intersection was a wood-frame, solidly built house, the home of the Rector family: Lester, Kathryn, sons Vernon and Ben, and a four-year-old daughter, Kathleen. Ben was away at college; Vernon had a date with a girl in Grandview and refused to call it off despite his mother's entreaties; he got in his car and headed east. A few minutes later, Lester, who had been outside watching the storm, came in, gathered Kathryn and little Kathleen and took them into the basement.

The tornado first hit the Ozanam Home for Boys; fortunately, the residents were huddled safely in the basement. The funnel then roared through Martin City, destroying or damaging virtually every home, business, and building in the town.

The core of the tornado passed a few hundred feet to the southeast of the Rector home. Kathryn's brother-in-law, Hoover Keister, owned K&K Filling Station, which was destroyed. But Hoover and his brother Jim were safe because they had taken shelter under the heaviest equipment they had, a large tractor. A steel girder fell on top of the tractor, but the tractor broke the girder's fall, leaving Hoover and Jim unharmed. As they crawled out from under the tractor they came face to face with . . . a cow.

The Martin City kindergarten graduation was in progress. As the tornado passed just south of the school, men were holding onto the men who were holding the school doors closed to keep the wind out.

The business district of Martin City was devastated, taking a direct hit from the tornado. Jess and Jim's Steak House was leveled, though the owner's parakeet—the only living creature in the building—somehow survived.

A birthday party was in progress in the basement of the Martin City Methodist Church. The participants were safe as the storm passed overhead. The building, however, was so heavily damaged it had to eventually be torn down and rebuilt.

Wrecked downtown Martin City, Missouri. The Rector home was just beyond the upper right edge of the photo. Reproduced with the permission of the *Kansas City Star*.

On his way to Grandview, Vernon Rector realized that the tornado was bearing down on him. He got out of his car and lay down in a ditch. The tornado practically passed right over him. The vibrating ground, the roar of the wind, the breaking glass, and the destruction were overwhelming. Though Vernon was unharmed, he had to walk part of the way home because the extensive debris on the road made driving impossible.

Northeast of Martin City at St. Catherine's School where the annual All-Sports Dinner was in progress, Monsignor Bernard Reiser was informed of the tornado and moved student athletes and guests into hallways as it passed one mile to the south. They prayed the rosary together while sheltering from the storm.

At the same time, Robert Jackson, the general manager of Debacker Chevrolet, decided to walk around the side of the dealership to view the tornado as it approached. His escape plan was to drive out of the way if the tornado got too close. After all, he was surrounded by cars. In the fifties, it was common for car dealers to keep keys in some of the most prominently displayed cars so that if a customer, on impulse, wanted to take a test drive, the customer and salesman could do so without delay.

As the tornado came into view, Jackson realized it was time to make his escape. He went to the closest car . . . no keys. A second one . . . no keys. There were no keys in any of the cars near the rear of the building. He then tried to take shelter in the building, but the back door of the dealership was locked. Jackson later described the experience in an article in *Weatherwise* magazine:

> [When the tornado was] about 150 feet away, I saw it pick up several cars and throw them 30 or 40 feet in the air, one of them sailing along with the wind around the outside of the funnel. The tornado came on across the street and our building began to fly apart before my eyes. A house to my right lifted off its foundation in one piece, then disintegrated, and wood flew everywhere.
>
> I saw this terrible thing coming; it was very black, just like mud. It had water, and mud, and everything in it, and it started clapping me on both sides and banging me behind this tree.

He grabbed the tree as the wind hit him. At times his feet were pointing nearly straight up and at times his feet were banging into the ground.

> Then, the blackness disappeared. The wind decreased to a near calm. I thought the tornado had actually gone over, so I stood up and looked around. I felt that I was about to be picked up into the storm, but at the same time, I felt "heavy."
>
> I realized I was looking up into the core of the tornado. I can't describe it, it was the most awesome thing. This column just [went] right straight into the heavens. It was bent toward the northeast up to about 200 to 300 feet, then straightened out and went straight up. I couldn't see anything whirling at all. I don't know how long I was inside the tornado, but I [realizing the backside of the tornado's winds were imminent] quickly lay back down again and grabbed hold of [the tree's] roots.
>
> Then, the back half hit me and it was black and smudgy again, dirt whipping through the air, automobiles hitting up and down. I remember that my shoes were about to come off my feet and I kept working my feet because I didn't want to lose my shoes for some crazy reason . . .

The drive-in theatre was next to go. Then the Copper Kettle Restaurant and Hickman Mills Bowl. The tornado continued northeast at a rate of 66 feet per second, chewing up homes, tossing cars, and stripping bark off trees.

From the national severe weather forecasting center:

7:44PM . . . THE MOST SEVERE THUNDERSTORM ACTIVITY IS NOW TO THE SOUTH AND SOUTH-EAST OF GREATER KANSAS CITY . . . REPORTS INDICATE THAT THE FUNNEL OBSERVED TWO

MILES WEST OF GRANDVIEW AIR FORCE BASE
WAS ALSO OBSERVED FROM THE GROUND AT
THE BASE. THERE ARE SEVERE THUNDER-
STORMS THROUGH THE AREA AND PRECAU-
TIONS SHOULD BE TAKEN BY EVERYONE.
PLEASE ASK RADIO LISTENERS NOT TO CALL
THE WEATHER BUREAU, EXCEPT TO REPORT
WINDSTORM DAMAGE OR THE SIGHTING OF A
FUNNEL CLOUD . . . WEATHER WARNING ISSUED
BY WEATHER BUREAU FORECAST CENTER
KANSAS CITY MO.

The air was filling with high-speed debris, one of the deadliest aspects of the storm. Pilots were reporting flying into debris at an altitude of 30,000 feet.

From the national severe weather forecasting center:

7:45CST MONDAY MAY 20 1957 . . . BRANIFF
PILOT REPORTS TORNADO ON GROUND NORTH
OF GRANDVIEW MOVING NORTHEAST. APPEARS
TO BE HEADED TOWARD EAST EDGE OF CITY.

At 7:47 p.m., time ran out for my neighborhood of Ruskin Heights. The tornado had arrived.

First to go were Tower School and Burke Elementary. With 250-mile-per-hour winds, the tornado was so strong it picked up a family trying to escape the storm by car, lifting the car about 100 feet, where it ricocheted off the top of the neighborhood's water tower. The car came to rest next to the Kansas City Southern Railroad tracks, with everyone inside dead except for one little girl.

The water tower itself remained standing.

The tornado crossed the railroad tracks and roared into the Ruskin Heights Shopping Center, flattening it on top of the people still inside.

When the tornado crossed Blue Ridge Road, the thoroughfare in front of the shopping center, it entered the dense area of prefab homes. The air filled with debris and screams. The tornado's north flank destroyed the new high school, leaving nothing but bare steel girders and the letters "RU—IN" where "Ruskin" had been before. The high school graduation rehearsal had ended earlier in the evening, but the adults remaining inside the school building were killed.

Aerial photograph of Ruskin Heights, May 21, 1957. Reproduced with permission of the *Kansas City Star*. My home was near Blue Ridge Road off the edge of the photo to the right.

The few basements in Ruskin were filled to overflowing with people. There was one report that a single small basement contained fifty people, some lying on top of each other.

Others, who had waited too long to take shelter, were overtaken by the tornado as they ran toward neighbors' homes or the nearby church. Most of them died instantly.

It was at this point that my dad came running into the house. From the local Weather Bureau office at the Municipal Airport:

```
AT 7:55PM RADAR INDICATES THAT THE TOR-
NADO OVER HOLMES PARK SHOULD MOVE
NORTHEASTWARD TO THE RAYTOWN VICINITY.
END. JFA
```

With numerous visual sightings of the tornado and myriad reports of devastation along its path, the prohibition against the field office using the word *tornado* was forgotten.

Dad had seen the green cloud turn toward our house, as the tornado was clearly visible from our front yard. While I probably saw the tornado through the picture window, I have no memory of it; I only remember the strange, dark green sky.

Once we were all safely in the basement, Dad turned on the radio, but we could hardly hear what the announcer was saying because of the near-constant crackle of the static caused by the intense lightning in the area.

From the local Weather Bureau office at the Municipal Airport:

```
FUNNEL OBSERVED AT 87TH AND RAYTOWN
ROAD AT 7:55PM
```

The scramble down the stairs, the indecipherable radio broad-cast, and the concern in my parents' faces all told me one thing: whatever was happening was bad. We had never before hunkered in the basement like this.

From the local Weather Bureau office at the Municipal Airport:

```
TO KEEP YOU UP TO DATE ON SEVERE WEATHER
IN THE AREA, THE MASSIVE THUNDERSTORM
WHICH HAS APPARENTLY MOVED ACROSS
```

THE SOUTHERN PORTION OF THE CITY IN
RECENT MINUTES IS NOW TO THE EAST OF THE
CITY AND THIS PARTICULAR STORM PROBABLY
POSES NO FURTHER THREAT TO THE CITY.

We stayed in the basement for what seemed like a very long time.
All I can remember is being scared.

From the local Weather Bureau office at the Municipal Airport:

SECOND REPORT JUST RECEIVED OF FUNNEL
OBSERVED JUST TO THE SOUTH OF RAYTOWN.
7:57PM . . . YET ANOTHER FUNNEL REPORTED
IN THE RAYTOWN AREA, OR RATHER A THIRD
REPORT OF THE FUNNEL BEING SIGHTED. 7:59.

While Mom and Dad were discussing what to do, the phone rang.
It was the Raytown, Missouri, police. Because my father's Ford dealership was in Raytown, they had our home number on file in case of
emergency. They asked Dad to open the building and allow them to
borrow station wagons for use as ambulances, due to the overwhelming number of injuries; early estimates were that over 1,000 people
were injured.

Our home had no structural damage, but we did have debris in the
front yard. We didn't know if there might be other tornadoes in the
area, but my father never hesitated. He headed out toward Raytown
in the direction of the recently departed tornado, which had passed
just a few miles south of the dealership. I have only a vague memory
of Dad's return, but I had rarely seen my mother so relieved. Dad's
example of bravery and public-spiritedness stayed with me the rest of
my life.

After leaving Ruskin, the tornado continued through mostly open
country until it reached a wide spot in the road known as Knobtown.

It then lifted on the southeast side of Raytown after traveling seventy-seven continuous miles on the ground.

It may have touched down several more times, as there were reports of damage from Unity Village, Independence, Liberty, and Ray County. If these reports were indeed from the Ruskin tornado, then its total path length was over ninety miles.

While the Fujita Tornado Scale, used by meteorologists to estimate the wind speeds associated with tornadoes, was not developed until 1971, the Ruskin Heights tornado was retroactively rated F5. F5 is the strongest category of tornado, with fewer than one-half of 1 percent of tornadoes reaching that level. A rain of debris was reported northeast of Kansas City. A cancelled check from Ruskin Heights was found in Ottumwa, Iowa, 180 miles to the northeast.

Given that type of force, even well-constructed and reinforced buildings such as the high school and shopping center were destroyed. Prefabricated homes didn't stand a chance. Meteorologists who do post-storm surveys privately say to each other, "You can tell it was an F5, because it cleaned up after itself."

The Ruskin Heights tornado was very tidy indeed. Concrete slab foundations were swept completely bare.

Some survivors' accounts illustrate the experience on a personal level:

> I had a baby cousin that died in that tornado. My aunt was paralyzed, and another cousin died later from complications he got from a blood disease due to the tornado. They lived catty-corner from the high school and were trying to get across to the Presbyterian church when the tornado hit. My baby cousin was taken from my aunt's arms and was found out by Lee's Summit. [Lee's Summit is four miles from Ruskin Heights.]

> Although I survived, my six-month-old sister, Linda Sue Stewart, passed away in the storm. She was ripped from my

mother's arm as my mom struggled to hold Linda, turn on a flashlight, and hold my hand, trying to get us to safety. The house imploded just as we took the first step to the basement.

We lost everything but the clothes we had on that night. My father also lost his job for some time, as Ruskin High School was destroyed as well. We were luckier than some though. We all survived. Even our collie dog in the backyard lived through it by crawling behind a brick BBQ oven that my dad had built. My parents have told many stories about the horrors of that night and the days following, as well as the kindness and charity of friends, neighbors, and strangers. My father spoke of that night and the acts of heroism, bravery, and superhuman strength of which he was a participant and witness until he passed away in September of 2005.

Three-year-old Bobbi Davis awoke in the hospital. She remembers nurses rubbing something in her eyes, and she asked them to do it some more because she wanted to see. Her father, mother, and sister were in one car, and her aunt and uncle in a second fleeing the tornado. They had just rounded the corner of Bennington and 113th. Her aunt and uncle's car just escaped the periphery of the tornado. Bobbi and her family were overtaken by the storm. She has no memory of the car becoming airborne, striking the water tower, or landing near the tracks. She was still in the hospital on her birthday, June 16, and they presented her a nurse-shaped cake.

The tornado killed forty-four people and officially injured 531 others. But this type of tornado, of F5 intensity, striking a densely populated area, typically killed more than 100. Why had so (relatively) few people died?

Earlier in the day, the Weather Bureau had issued a tornado watch, raising public awareness of a potential threat. The continuous coverage and Audsley's direct, to-the-point messages clearly saved lives.

The Ruskin Heights tornado was the first time the fledgling Weather Bureau tornado forecast and tentative warning system had had any degree of success.

Almost as soon as the tornado was gone, sightseers flooded into the area. There were reports of looting. The Missouri National Guard was called in and deployed. Residents had to carry a special tag to get to areas inside the disaster area where martial law—civilians put under military rule—was in effect.

We hung the three ID tags from our car's rearview mirror to get past the checkpoint near our home. While the tags were intended to allow us to get to our home, they also allowed us free access to the tornado devastation. The next day, we went out to see what had happened.

The events of the evening of May 20 didn't make sense to me under the sunny sky of May 21. But as we drove south on Blue Ridge Road toward Ruskin, the impact of the tornado began to come into focus. We encountered more and more damage, and the damage became more severe.

After making some detours onto side streets, we arrived in the center of the damage path: trees completely stripped of leaves and bark, twisted bare girders of the new Ruskin High School, and the homes that were *gone*, with nothing but bare concrete slabs as proof of their existence.

Until the 2007 Greensburg, Kansas, tornado—fifty years to the month after Ruskin Heights—I've never seen destruction quite as complete. Even today, I don't have words that adequately convey the Ruskin Heights scene. The tornado's path was so long and so wide that in the middle of the damage path nothing but rubble was visible in all directions.

Suddenly, the events of the night before made sense. And, even at the age of five, the thought that would change my entire life went through my mind: *Anything that could do all of this has to be pretty interesting.*

From that moment on, I knew that I wanted to study whatever had done "all of this."

Most meteorologists can point to a single event, usually a storm they experienced between the ages of five and nine, which initiated their lifelong interest in weather. In fact, the Ruskin tornado inspired not just me, but at least two other people to become meteorologists.

* * *

The ten-year period from 1947 to 1957, culminating with the Ruskin Heights event, featured an unusually high number of extremely destructive tornadoes. The worst of them was in an obscure part of the High Plains, and the story of how the Weather Bureau was dragged, kicking and screaming, into the tornado business begins in the Texas Panhandle.

CHAPTER TWO

NO ONE EVER KNEW
IT WAS COMING

FOR SEVENTY YEARS THE WEATHER BUREAU emphatically discouraged the ambitions of one meteorologist after another who wanted to study or attempt to forecast tornadoes. The story of how the modern tornado warning system began starts in a remote area of the southern High Plains, just after World War II in April 1947.

In 1946, a song written by Bobby Troup, "Route 66," captured the public's imagination. U.S. Highway 66 was a two-lane blacktop road across Oklahoma and the Texas Panhandle that was the primary U.S. highway from Chicago to Los Angeles. However, the main *economic* artery in the Texas Panhandle and northwest Oklahoma in 1947 wasn't a highway at all.

In the 1910s, the Atchison, Topeka, and Santa Fe Railway extended its main line from eastern Kansas southwest across the Texas Panhandle. Both passenger and freight trains traveled at much higher speeds (with passenger trains reaching 80 miles per hour in places) than the posted speed limit of 35 miles per hour on much of high-ways U.S. 66 and U.S. 83, or the other narrow roads spanning this vast expanse of territory. Railroads were major economic attractions, so jobs and towns tended to develop along their main lines. Heading northeast from Amarillo, Texas, spaced every few miles along the railroad, were the towns of Panhandle, White Deer, Pampa, Miami, Canadian, and Glazier.

The Texas Panhandle was—and is—lightly populated; in fact, the federal government later located the United States' major nuclear weapons facility, called Pantex, along the railroad northeast of Amarillo. In the 1980s, Pantex manufactured and shipped nuclear weapons to military facilities around the country on what came to be known as the "White Train." All of the train's cars were painted white so they would be instantly recognizable when the train's progress was monitored by air.

April 9, 1947, dawned misty and gloomy across northwest Oklahoma and the Panhandle. The weather was not the warm, muggy weather usually associated with tornadoes. The few weather stations in the area reported low clouds, drizzle, and cool temperatures in the low fifties. Dense fog was reported in many areas, with visibilities near zero. Even longtime residents would not have walked outside that morning and thought "tornado": it was the wrong kind of weather—too damp and cool, and not muggy enough.

The unusually dense fog was an inconvenience. But another development was evident only to those with barometers—the pressure was falling rapidly. It went from 29.78 inches at 10 a.m. to 29.34 inches in less than eight hours, which constituted a huge drop. The Weather Bureau's forecast that morning, though, was for nothing more ominous than "thunderstorms with heavy rains."

A few hours later, all of this foggy moisture was diverted skyward into towering cloud columns. A farmer surveying his deep green winter wheat at the end of the day peered through new breaks in the low clouds east of Amarillo, and if he stared at just the right spot in the sky, a boiling cumulus cloud caught the sun and created a near-blinding contrast to the leaden lower clouds.

The simple act of water vapor transforming into liquid cloud droplets (condensation) produces a staggering amount of energy. For each gram of water vapor that turns to liquid water, 540 calories of energy are released. This might not seem like much, until you realize that a supercell thunderstorm (the most severe variety of thunderstorm) is formed from *trillions* of grams of water vapor, each one releasing 540 calories.

Drawing on calculations by *USA Today*'s meteorologist Bob Swanson, an ordinary thunderstorm produces about as much energy as a half-megaton nuclear weapon. A supercell thunderstorm—the kind that was forming just northeast of Amarillo—produces many *megatons* of energy. When that energy is concentrated in a violent tornado, the resulting destruction is comparable to that caused by an atomic bomb. So in some ironic way, the birthplace of this particular supercell thunderstorm—one of the strongest in history—was appropriate: it was above the site of what would become the Pantex nuclear weapons facility.

Over the next five hours, it would look as if an airplane, flying northeast along the railroad tracks, dropped a string of atomic bombs on the Texas Panhandle.

* * *

In order to understand what was so extraordinary about the weather of April 9, 1947, we have to understand a little of what we know today about how ferocious tornadoes—the ones with the highest wind speeds—develop.

We know that air can move parallel to the ground (what we call wind), or it can move up (updraft) or down (downdraft). Most rain showers or thunderstorms have brief multiple updrafts that carry the moisture needed to initiate and sustain the precipitation. A supercell thunderstorm has a single long-lived updraft that rotates, that is, the wind spirals in a circle as it ascends. And it is via a supercell that the most vicious type of tornado is formed.

It's hard to overstate the sheer power of these supercell storms. The air rises in the updraft at speeds that can exceed 100 miles per hour. In 2007, a German paraglider was sucked into the updraft of an Australian supercell thunderstorm. The instrumentation she carried revealed she was lofted to an altitude of 32,725 feet. The paraglider survived, but she experienced severe frostbite (temperatures were as cold as −50°F) and received bruises from hailstones the size of oranges.

A second paraglider was sucked into the same storm and perished. His body was found forty-seven miles away.

On August 6, 1966, Braniff Airways Flight 250 unknowingly flew into the updraft of a supercell thunderstorm. The plane was lifted so quickly that the tail fin and the right horizontal stabilizer (one of the small wings at the back of a plane) broke off. The aircraft experienced one of the rarest and most nightmarish events in aviation, an in-flight breakup. It fell onto a farm near Falls City, Nebraska. All forty-four people aboard were killed.

Given the right conditions, a violent tornado can spin up in the rotating updraft of a supercell storm. Fortunately, a supercell is a meteorologically rare event. In order for a supercell thunderstorm to form, four things must occur in approximately the same place at the same time:

1. The air must be unstable (meaning, to meteorologists, that if a cubic foot of air starts to rise, it will continue to rise, which will help form an updraft). The air on April 9, 1947, in the Texas Panhandle was extraordinarily unstable.

2. The air over the region as a whole must be rising. This occurs with strengthening broadscale (i.e., covering the geographic

area of, say, Illinois and Indiana) low pressure. The rapidly falling pressure over the area that day was a danger sign that, even though temperatures were only in the fifties (usually too cool for tornadoes to form), the weather was not as it seemed. In ways invisible to the eye, the environment was changing from one unfavorable for tornadoes to one that was extremely favorable.

3. Within the broad area of ascending air, there must be a more concentrated area of rising air (i.e., air that is displaced upward). If for some reason the winds converge in one area, the air will be forced to rise in the same way a sheet of paper lying on a desk will rise in the middle if you press the ends of the paper together. Winds converge in the presence of a front or other wind-shift line. The tornadic supercell forms along or just east of the wind shift.

4. The final ingredient for development of a supercell is the profile of winds as one ascends in the atmosphere. Wind directions must "veer" with height—in other words, turn in a clockwise direction as viewed from above. For example, the winds may blow from the southeast at ground level, from the southwest at 5,000 feet, and from the west at 20,000 feet, and speeds must accelerate with height (that is, wind speeds must be greater at 30,000 feet than at 5,000 feet). This atmospherically infrequent state of affairs is necessary to evacuate the rapidly rising air of the updraft; otherwise, the storm would quickly collapse of its own weight. It is also necessary to sustain rotation in the supercell's updraft over a relatively long period of time. Ordinary thunderstorms typically last twenty minutes or so. Supercells often last hours.

None of this was known in 1947. Meteorologists would not understand supercell formation for decades.

All that residents of the Texas Panhandle and western Oklahoma knew was that there was a moist, gusty southeast wind that blew throughout the day. In the afternoon, around four o'clock, the fog started to lift, the overcast turned to broken clouds, and temperatures began to rise. In this context, the rising temperatures were bad: warmer temperatures near the ground made the air more unstable. The weather station at Pampa, Texas, reported a temperature of 58°F at 3:30 p.m., the time of day when temperatures usually peak. Instead, on April 9, the temperatures started rising rapidly, reaching 65°F by 6:30 p.m. and rising to 69°F by 7:28 p.m. (a time of day when temperatures should be falling). The air was becoming more unstable, almost by the minute.

Condition number one for supercell formation was met.

Panhandle farmers glancing up that afternoon first noticed the low scud clouds, about 800 feet above the ground, racing from southeast to northwest. It seemed the weather was improving. If they looked closely, they could see cumulus clouds a couple of thousand feet higher that were moving from straight south to straight north (condition number four). The barometers continued to unwind as pressures kept dropping (condition number two).

Metaphorically, the hammer was cocked. Now all the atmosphere needed was a nudge.

The weather maps from that evening show two wind-shift lines. One, in the western Texas Panhandle, was the east edge of dry air from New Mexico carrying blowing dust. The second was more subtle: a shift of wind from south to southeast near Amarillo.

At 6:04 p.m., the Weather Bureau station at the Amarillo airport recorded partial sunshine and a subtle wind shift from south-southeast to straight south. The humidity went down rapidly. The temperature shot up five degrees, even though it was late in the day for that rapid a temperature increase. If you squinted, you could see the blowing dust from New Mexico that tinted the sky on the western horizon a reddish brown color.

But the more important action was just east of the city.

The wind shift that moved across Amarillo provided the final ingredient (condition three) for a supercell. The winds east of the city were blowing from southeast to northwest, while winds over the city were blowing from straight south to straight north.

This *convergence* of winds at ground level caused the colliding air to start to rise. Once this upward air motion had begun, the off-the-chart instability caused the air to accelerate upward. The air was like an inverted waterfall, rushing upward, unseen and unheard.

A supercell thunderstorm was born.

Although there were no measurements taken in 1947, it is certain that this supercell thunderstorm towered at least ten, and probably twelve or more, miles into the atmosphere. Cirrus clouds spread rapidly northeast ahead of the main cloud mass, caught up in the jet stream's winds. The storm's rotating updraft got stronger with time, and patterns of the striations in the vertical wall of the cloud may have resembled the alternating stripes of a barber's pole. From a hundred miles to the north or south, the shape of the storm would have resembled a blacksmith's anvil in the sky.

Any pilot would know instinctively to give this storm a wide berth.

As bad luck would have it, the winds from two to five miles above the ground were unusually strong, meaning that thunderstorms caught up in them would move faster than normal. The still misty weather east of the wind-shift line reduced visibility, and if a tornado developed, the thunderstorm's faster-than-average movement of 45 miles per hour made it highly unlikely that word could get out about the storm before telephone and telegraph lines were destroyed. Since it was early April, the sun was low on the horizon and would soon set. Darkness and fog would make advance sighting and notification impossible.

The number of unfortunate circumstances was mounting: The strong supercell had formed when the four ingredients came together, and the storm was moving faster than normal. These two factors were dangerous enough. But the third unfortunate circumstance set the stage for the incredible disaster that was to follow: the upper atmospheric winds were just right to move the thunderstorm northeast along the Santa Fe tracks and toward the series of towns situated along them. A mere five-mile displacement to the north or south would have caused the tornado to move through largely unpopulated areas.

As the thunderstorm began its northeast journey, the base of the cloud lowered and began to rotate visibly. This would have been a clear sign of danger had anyone been looking or known what to look for.

Given these circumstances today, a tornado watch would have already been in effect. With the supercell's rapid development and intensity displayed by radar and weather satellite, undoubtedly a severe thunderstorm warning would be in effect, bringing out trained storm spotters, ready to report rotation or a funnel so a tornado warning could be issued.

There were no storm spotters in the Texas Panhandle in 1947. No one was looking for the danger signs that a tornado might be forming. For 221 miles, this tornado would be a tragic surprise. It was a tornadic "perfect storm."

Around 5:48 p.m., a few miles west of White Deer, the thunderstorm had developed at least 100 miles per hour of rotation aloft. The rain and hail descending on the supercell's back edge was widespread enough to constrict the width of the rotating air near the cloud's base.

The constriction causes the air's speed of rotation to increase in the same way a spinning ice skater spins even faster when she pulls in her arms. The rotation stretched vertically up into the updraft and down toward the ground.

A tornado was born.

The tornado first kicked up dust a half-mile south of White Deer, Texas. The path of the tornado and the railroad tracks were intertwined: the first victim was an eastbound Santa Fe freight train at White Deer. The tornado overtook the train, derailing nineteen cars and three cabooses, and injuring two people. As sometimes happens with supercell thunderstorms, the first tornado was relatively small; it turned north and dissipated after a few miles.

To the south, a second, far more sinister tornado formed and began moving northeast.

Next in line was the town of Pampa, fourteen track miles to the northeast of White Deer. Pampa had a small airport. The Federal Aviation Administration's (FAA) weather observer at the airport was charged with observing and recording the weather conditions as they related to aviation. His first job was to go outside and look at the sky to observe the clouds and visibility. He then would come inside to read the instruments that measured the air temperature, the dew point, winds, and pressure, and he would write his observations onto a form.

You might think the FAA's weather observer would have stopped the presses and sent his findings immediately. You might also think he would have telephoned the police, sheriff, and other authorities to relay this crucial information. Unfortunately, this was not the case. Because of the restrictions imposed by the Weather Bureau and the FAA on their employees, there was no institutional urgency to disseminate the information.

On this night, the next opportunity to transmit the vital weather information was the 6:20 p.m. transmission. Every twenty minutes, the regional teletype operator stopped the circuit to let the FAA field offices transmit "special" observations of important changes in the weather near the airport. Most Tornado Alley residents live their entire lives without seeing a tornado. The Pampa weather observer was likely counting the seconds so he could get this nearly one-of-a-kind observation out to the world.

When the field office poll began, there was a pause in the traffic on the teletype circuit. Then, the three-letter airways code for each station was transmitted, followed by a pause of a second or two to see if that weather station would begin transmitting.

PPA USP 1810 E30⊕ 10 TORNADO \29+ TORNADO
N OF STN MOVING NE

This bulletin states that an urgent special observation (USP) was
recorded at the Pampa, Texas, airport (PPA) at 6:10 p.m. CST. Estimated ceiling (the height above the ground of the base of the clouds) is
3,000 feet with overcast conditions. Visibility is ten miles. Tornado in
progress within sight of the weather station. Wind southeast gusting
above 29 knots. The tornado is north of the weather station moving
toward the northeast.

After this transmission had been sent, all the airports on that circuit would have instantly known there had been a tornado at Pampa
ten minutes earlier. And in another few minutes, airports and Weather
Bureau offices all across the United States would know about the
Pampa tornado. The only people who *wouldn't* know? The residents
of the northeast Texas Panhandle and northwest Oklahoma.

Why? There are two reasons: The first is a ridiculous bureaucratic
myopia that exists in the FAA to this day. The FAA believes its tax-
payer-funded meteorological mission is to serve aviation *exclusively*.
So radio stations and others could not—and cannot—subscribe to the
teletypes that transmitted data from those circuits. A pilot in St. Louis
or Chicago would have been alerted to the Pampa tornado; the local
radio station would not have been.

And second, as we have said, at that time the Weather Bureau tied
its own hands by forbidding any mention of tornadoes in its forecasts
and warnings.

So the Bureau's aviation forecasters may well have known about
the Pampa tornado (if they happened to be monitoring the Service A
circuit at the time the message went across), but they were forbidden
to do anything about it. The Weather Bureau thought panic would
ensue if it issued a tornado warning, even if a tornado was on the
ground posing an imminent threat to life and property.

As the last few inches of the Service A message tape exited the tape reader and fell into the wastebasket, the only chance to warn people northeast of Pampa was gone.

Pampa itself, as it turned out, received only a glancing blow, as did the town of Miami. But the now mile-and-a-half-wide tornado attacked Glazier, killing seventeen people and razing the town: only one building was left standing. Early the next morning, a TWA pilot attempting to use Glazier for navigation was temporarily disoriented when he couldn't find the town.

At this point, the Pampa tornado had covered sixty-eight track miles.

Like dominoes, one town after another fell victim to this tornado as it moved northeast along the Santa Fe. Sixteen miles northeast of Glazier was Higgins, population 750. Fifty-one people died there. Only four buildings were left even partially standing, and those were destroyed minutes later when gas lines ignited, causing a huge fire. By the time the fire was extinguished, not a single inhabitable building was left.

And it wasn't just towns that were hit: accounts vary, but it seems that two more trains were overtaken by the tornado and thrown off the tracks. Already sixty-nine people were dead and more than 200 injured.

But this tornado was just getting warmed up.

A mile east of Higgins is the Oklahoma border, where the railroad track turns north. At this point, the tornado and track parted company for the first time. The tornado made its way northeast across the low rolling hills of Ellis County, Oklahoma, where scattered farms and ranches were in its path. Sixty of them were destroyed, eight people were killed, and forty-two others were injured.

The classic supercell drops its heaviest rain and hail north of the path of the tornado (the tornado usually occurs on the southwest edge of a northeast-moving storm). This supercell was no exception, as half-inch hail began falling at the tiny Gage, Oklahoma, airport at 7:28 p.m. The winds became somewhat more east-southeasterly and gusted above 40 miles per hour. The barometric pressure at Gage, which had been dropping fast all day, went into a free-fall—so much so that the weather observer twice noted the rapid pressure drop. He also noted a "vivid" display of "continuous" lightning.

The tornado crossed out of Ellis County and passed south of the town of Tangier. Another person died. While the nine Oklahoma fatalities were a tragedy, the orientation of the track, and the towns along it, was such that the tornado missed the bulk of the population of Ellis County.

Just east of where the Santa Fe railway crosses Sand Creek, the railroad track takes a turn back to the east, and that's where it reunited with the tornado: at the largest town in northwest Oklahoma, Woodward.

Because this storm was moving quickly, it was dark, and the tornado was more than a mile and a half wide, it's likely that no one realized what was about to happen. Even if someone had been outside looking for it, there was no funnel cloud as we normally think of them. There was just a huge, dirty, cloudy wall of 250-plus mile-per-hour winds.

Words cannot describe the manner in which a tornado of this intensity destroys a town. The only expression that comes close is "buzz saw." The air fills with debris flying at over 200 miles per hour in the same way that sawdust spins out of a whirling blade. Nail-laden boards, glass shards, clumps of asphalt torn from roads, sheets of jagged metal, and automobile chassis fly through the air just like sawdust flies from the blade. Tall, mature trees are transformed into bald, barkless six- to ten-foot-high spears.

Survival in an F5 tornado, unless you are securely underground and out of the direct line of fire, is simply a matter of luck. And even if you are safely in a basement, survival is not 100 percent assured.

A hundred city blocks were destroyed in mere moments. The death toll was 107 people, with more than a thousand injured. Debris from Woodward was swept up and out the virtual chimney of the supercell's 100 mile-per-hour updraft and didn't fall until it had traveled sixty miles across the border into Kansas.

Though citizen-rescuers were on the scene within fifteen minutes, Woodward was cut off from the outside world. There was no way to call for help quickly.

Oklahoma Gas and Electric Company had a facility in Woodward in the direct path of the tornado. Irwin Walker realized what was happening, and rather than trying to find shelter within the building, his final act was to throw the switch that cut off electricity to the town, de-energizing the hundreds of lines that were on the ground or blowing through the air, saving lives and preventing the fires that would ordinarily have been sparked with live wires flying everywhere.

When workers got to the phone company building, there were no circuits functioning anywhere in town. Once the storm passed, a tele phone worker's test phone was used to call Oklahoma City to request help. Phone company workers began restoring service as quickly as possible, and operators began trying to complete rescue calls.

There is a unique smell to an area that has been devastated by a tornado. Pulverized, rain-soaked building materials exposed to air for the first time in years or decades, broken pipes spouting natural gas, household and industrial chemicals released and exposed, and—worst of all—rotting biological materials combine to foul the air.

The debris is heartbreaking. Perhaps it is because the heavier objects (iron girders, automobile frames, etc.) fall to the ground first, there always seems to be a sprinkling of children's toys on top of the debris. A well-worn, mud-spattered doll, a rain-soaked baseball glove, a large broken plastic sign from a wedding store with just one word intact: "Memories."

It isn't just human and animal life that is affected: normally well-tended lawns turn brown almost instantly. Just as unwatered grass can turn brown after several days of hot, windy weather, the 200-plus mile-per-hour winds of a violent tornado can deliver the equivalent of weeks' worth of battering wind in a matter of seconds. The path of a strong tornado can often be tracked by air over the grasslands of the Midwest by the brown path it leaves.

Given the unfamiliar smell, the surreal vegetation, and the fact that buildings, landmarks, and street signs are gone, even longtime residents of a devastated area can be completely disoriented.

The citizen-rescuers in Woodward had to deal with all of this—plus the horror of finding more than a hundred bodies and corresponding body parts strewn throughout the wreckage.

Rescuers wish to be sensitive and respectful to the deceased men and women and their remains, but they also want to move as quickly as possible so the injured can be found and care rendered. Trying to balance the two objectives leads to almost unimaginable stress, leaving an indelible image in the minds of the rescuers. Some are traumatized by the experience, especially citizen-rescuers who have never been in this or a similar situation. Nightmares tend to plague rescuers, and mental health issues centering around survivor's guilt are likely to develop.

Meanwhile, the tornado was continuing on its northeast path. The railroad tracks go east out of Woodward before they resume their northeast orientation. The tornado and railroad track parted company for the last time. The massive storm passed just west of the railroad towns of Alva and Avard, injuring thirty more people.

The supercell crossed the border into Kansas west of the town of Medicine Lodge, in a sparsely populated area known as the Gypsum Hills, an area where cattle roam the fenceless open range. The Gyp Hills in 1947 was not only nearly devoid of people, it was without telegraph or many telephones.

Not much is known about the tornado's path through the Gyp Hills. The tornado and/or high winds associated with the now slowly weakening supercell continued to move north-northeast into Kansas and caused damage in and near the towns of Hazelton, Sharon, Zenda, Nashville, and several other small communities, with lighter scattered damage over the area.

After more than five hours, the last damage was reported near the tiny town of St. Leo, in western Kingman County, roughly forty miles west of Wichita. This tornadic supercell, and the tornadoes it spawned, killed 181 people and injured over 1,500.

And for its entire 221-mile path, no one ever knew it was coming.

* * *

After enduring more than two decades of unprecedented death and destruction from tornadoes, the public and the press alike began letting politicians know that the nation that won World War II and created the atomic bomb should be able to do something about tornadoes.

West of Amarillo, Route 66 and the Santa Fe tracks parallel each other. Beyond Winona, Kingman, and Barstow, both highway and tracks reach their destination of Southern California. Little did anyone dream that the person who would provide the first critical piece of the tornado puzzle grew up at the western end of Route 66.

And he had never given tornadoes a moment's thought.

CHAPTER THREE

"NICE PEOPLE, BUT ODD"

IN 1884, JOHN PARK FINLEY OF THE U.S. ARMY SIGNAL Corps published a scientific paper in the *American Meteorological Journal* and *Science* concerning an experimental tornado forecasting system: based on gathered data, he believed that tornadoes could be forecasted, and he presented six rules for doing so.

Finley's method met with mixed success, but the end result was that for the first time, the Signal Corps (the predecessor of the U.S. Weather Bureau) created a special category of warning for violent storms.

Unfortunately, the Signal Corps was involved in a number of scandals in the mid-1880s; Finley did not play internal politics well and was kicked out of weather forecasting, his tornado forecast methods ignored or forgotten. In 1887, the *Report of the Chief Signal Officer* noted that "it is believed that the harm done by such a prediction would eventually be greater than that which results from the tornado itself." As we noted in previous chapters, the Weather Bureau made a significant attempt to downplay the importance of tornadoes.

Mark W. Harrington, professor of astronomy and director of the observatory at the University of Michigan, took control of the new civilian U.S. Weather Bureau in 1891. But a lengthy feud between Harrington and the secretary of agriculture led to the appointment of Willis L. Moore, a severe critic of the Signal Corps' tornado reporters and, to his mind, their questionable methods of collecting data, as Bureau chief in 1895.

The new chief did not propose adoption of any new tornado counting or forecasting methods; instead, he ordered a review of all tornado reports from 1889 to 1896 with a view to adjusting the statistics to reflect only death and destruction by actual tornadoes, not all windstorms. Moore believed that "in almost all cases of great disaster there is a pronounced tendency to exaggerate the actual facts," and blamed journalists and tornado insurance companies for inflating statistics. Unquestionably, the United States had suffered great loss of life from tornadoes during the eight years under consideration, but Moore hoped to dispel the idea that the frequency or severity of tornadoes was increasing.

In response to an editorial in the *Chicago Tribune* questioning the dearth of tornado warnings, noted meteorologist Cleveland Abbe listed four reasons for not issuing them:

1. Lack of knowledge as to the tornado's direction of movement and the possibility the warning would not cover the correct geographic area;

2. No ability to forecast the duration of tornadoes;

3. Telephone operators (afraid for their safety) would fail to forward the warnings;

4. Three-quarters of all tornadoes would go undetected.

Abbe summarized his arguments by saying the Weather Bureau had "no right to issue numerous erroneous alarms" and the "unnecessary fright would be worse than the storms themselves." Note that at this time the Weather Bureau failed to issue public-opinion surveys or make any attempts to learn what the public wanted.

This line of anti-warning thinking continued. Weather Bureau regulations, reissued multiple times in subsequent years and decades, flatly proclaimed "forecasts of tornadoes are prohibited." Not only were forecasts and warnings forbidden, there was virtually no meteorological research on the topic for decades.

So when the great Tri-State Tornado of March 18, 1925, killed 689 people, the unfortunate residents of Missouri, Illinois, and Indiana were sitting ducks. So were the 216 white residents of Tupelo, Mississippi, who died in the tornado of April 5, 1936, and the 203 white residents of Gainesville, Georgia, who died the following morning. (Black residents were not included in statistics until the 1950s, even though in the Tupelo tornado they were the most affected. Hundreds of black residents may have been killed, bringing the unofficial death toll to 400 or more.) Those two tornadoes alone injured more than 2,500 people.

According to Marlene Bradford in *Historical Roots of Modern Tornado Forecasts and Warnings*, "a systematic approach to tornado forecasting and warnings was as nonexistent in 1940 as it had been in 1870." That attitude finally changed during the third week of May 1948, but it wasn't the Weather Bureau that changed it.

* * *

Like most people growing up in Southern California, young Robert Miller thought little about weather. His interests included math and science, and his career aspiration was to teach high school algebra.

He initially enrolled in a teacher's college and then transferred to Occidental College in 1941 to study math and physics. When World War II broke out, he enlisted in the Army Air Corps; when he'd had enough of Kitchen Patrol (cleaning the mess hall after a meal, better known as "KP"), he enrolled in the army's weather forecaster school in Grand Rapids, Michigan.

According to an article in the *Bulletin of the American Meteorological Society*, there were fewer than 400 weather forecasters in the United States at the start of the war. The Air Corps' need for forecasters was so overwhelming that it developed a nine-month forecasting course; Miller graduated from the corps' first class. With no real-world forecasting experience, he was assigned to March Air Force Base in California, and his first forecast, for pilots flying to San Diego, was a complete bust; if a review by a second forecaster had not occurred, pilots might have flown into dense fog with zero visibility and likely suffered disastrous results.

Within weeks, Miller began a series of stints at various bases as part of his Air Corps assignments. During most of this period, he was forecasting *tropical* weather, quite a different task from forecasting weather in the continental United States.

At the end of World War II, Miller was assigned to Fort Benning, Georgia, where he cleverly adapted an oceanography technique to atmospheric analysis. He was able to draw maps of the surface weather at various altitudes—4000-feet weather, 10,000-feet weather, and 18,000-feet weather—and view them simultaneously.

This three-dimensional picture of the weather enabled Miller to not only make better forecasts but also gain insight into how the atmosphere behaves as it does. The ability to think in three dimensions is what often separates good forecasters from great forecasters. The weather we experience near the ground is the product not only of the temperature, winds, humidity, and other factors near the ground, but also of what is occurring from one to six miles above the ground.

On March 1, 1948, Miller arrived at Tinker Air Force Base, Oklahoma. Tinker is located on the southeast side of Oklahoma City, immediately adjacent to the suburbs of Midwest City and Del City. Central Oklahoma is very flat with relatively few trees and almost no haze or smog, unlike the other cities where Miller had lived. His schooling and experiences to date had not prepared him for the rapid changes in thunderstorm weather that occur in the Great Plains.

The events that unfolded from March 20 through March 25, 1948, are unprecedented in meteorology, and perhaps, in all of science. What follows is a mix of my description along with Colonel Miller's own words as transcribed by Charlie Crisp of the National Oceanic and Atmospheric Administration's (NOAA) National Severe Storms Laboratory.

MARCH 20, 1948

The evening of March 20, I was rudely awakened to the sometimes vicious vagaries of Mother Nature. There were two of us on shift that night. My backup forecaster was a Staff Sergeant, also new to the Tinker Weather Station. In course of idle conversation, we found we had much in common—we were both from sunny Southern California and had no weather experience in the Midwest portion of the United States. We analyzed the latest surface weather maps and upper charts and arrived at the sage conclusion that except for moderately gusty surface winds, we were in for a dry and dull night.

Miller and the sergeant didn't know, because of an incorrectly analyzed weather map and their own inexperience, that strong thunderstorms rather than "dull" weather were in the offing.

Shortly after 9:00 p.m., weather stations to our west and southwest began reporting lightning and by 9:30 p.m. thunderstorms were in progress and, to our surprise, detectable only twenty miles to the southwest of the Base. Even on our crotchety old radar the leading thunderstorm cells looked vicious and were moving very fast. The Sergeant began typing up a warning for thunderstorms accompanied by stronger gusts even though we were too late to alert the Base and secure the aircraft.

At 9:52 p.m., the squall line moved across Will Rogers Airport seven miles to our west-southwest. To our horror they reported a heavy thunderstorm with winds gusting to 92 miles per hour and, worst of all, at the end of the message, "TORNADO SOUTH ON GROUND MOVING NE." We had it for certain! We could only pray that this storm would change course and move southeast.

There was no such miracle and at 10:00 p.m. the large tornado, visible in a vivid background of continuous lightning and accompanied by crashing thunder, began moving from the southwest to northeast across the Base. We watched it, not really believing, as it passed just east of the large hangars and the operations building where we crouched in near panic. Suddenly the glass in the control tower to our right succumbed to the pressure differential caused by the vortex, and all the glass shattered. The control tower personnel were badly cut. They had not abandoned the tower despite the 78 mile-an-hour winds around the outer fringe of the tornado. Seconds later the Operation Building's large window blasted outward into the parking area. Debris filled the air. Then, suddenly, the churning funnel lifted and dissipated over the northeast edge of the Base.

MARCH 21, 1948

The Air Force moved quickly. Twelve hours later, an investigative board flew in to Tinker from Washington, D.C., for a hearing on the event.

Major Fawbush (E. J. to me) and I waited our turn "on the grill" with considerable trepidation. I was especially tense, having performed in such an abysmal manner the previous evening. It really did not seem fair that a bright young forecaster, native to an area where a mild thunderstorm was considered a holiday event that caused people to run outside and gesticulate skyward mouthing such phrases as "golly" and "wow," should be thrust into an area subject to such miserable phenomena. The time came and we were ushered into the room. We snapped to attention with E. J. advising the board, in a garrulous voice, "Major Fawbush and Capt. Miller, reporting as directed."

As Miller put it, "the interrogation began." The board of inquiry asked about the events of the previous evening, the difficulty with forecasting tornadoes when no recognized techniques existed to do so, and the resulting reluctance to issue a warning of a tornado.

The board reached its decision early that afternoon. They decided that "due to the nature of the storm" it was not forecastable given the present "state of the art" and that "it was an act of God"—which it most certainly was. They recommended that the meteorological community consider efforts to determine a method of alerting the public to these storms and urged Base Commanders to develop safety precautions to minimize personnel and property losses in violent storms.

That afternoon the Commanding General of the Oklahoma City Air Material Area, Fred S. Borum, directed the Air Weather Service to have the Tinker Base Weather Station (under the command of Major Ernest J. Fawbush) investigate the feasibility of forecasting tornado-producing thunderstorms. Major Fawbush had been interested for some years in such storms, and since I had become "most interested" overnight, I was most fortunate in being selected to aid in the investigations.

MARCH 22–24, 1948

Fawbush and Miller set to work. Over the next three days they reviewed the weather maps pertinent to the Tinker tornado as well as those for a number of other tornadoes. They combined what they learned from the maps with some recent research published by three Weather Bureau meteorologists.

Fawbush and Miller found that tornadoes usually occurred in areas of warm, moist air with relatively strong winds aloft. They found several configurations or patterns of weather systems. This allowed them (and future severe weather forecasters) to use a technique known as "pattern recognition": Does today's location and intensity of fronts, pressure systems, areas of high humidity, and so on, resemble a composite of previous tornadic events?

Eventually, these patterns would be given names (i.e., the worst tornadoes tended to form in Miller Type 1 patterns).

> The problem faced by the forecaster was to consider the current surface and upper air data, determine the existence of these parameters or the probability of their development, and then project the parameters in space and time in order to issue the "tornado threat area" with a reasonable degree of confidence and lead time.

Lead time is the difference between when a warning is issued and when the storm occurs. It's a concept meteorologists use to determine the utility of a warning or forecast.

> Such a detailed forecast procedure was time and labor consuming and required intensive and specialized analysis.

MARCH 25, 1948

After their three-day period of intense study, Fawbush and Miller thought they would return to their normal routine.

Mother Nature had other ideas.

> On the morning weather charts of the 25th of March 1948,
> just five days after the Tinker storm, we noted a great simi-
> larity between the charts of the 20th and the 25th. After
> analyzing the surface and upper-air data, a prognostic chart
> was prepared for 6:00 p.m. local time showing the expected
> position of the various critical parameters. This chart resulted
> in the somewhat unsettling conclusion that central Okla-
> homa would be in the primary tornado threat area by late
> afternoon and early evening.

The base commander, General Borum, was an amateur meteorologist. According to Miller, he was "most knowledgeable" about
weather and was highly proficient in the operation of the Base
weather radar. When Fawbush and Miller realized that there was a
possibility of a second tornado, they notified the General and briefed
him on the situation.

> He [Borum] digested what we had told him and asked, "Are
> you planning to issue a tornado forecast for Tinker?"
>
> There was a period of uneasy quiet until E. J. spoke up.
> "Well, it certainly looks like the 20th, right, Bob?"
>
> Oh, great! I wanted to turn and ask my Sergeant friend
> the same question, but he was not on shift.
>
> I replied, "Yes, E. J., it certainly looks like it did on the
> 20th."
>
> After hearing these helpful observations, the General
> asked what we believed the critical time would be and
> received a useful answer this time—5:00 to 6:00 p.m. The
> General then decided we should issue a forecast for heavy
> thunderstorms during that period. He patiently explained
> that such a move would serve to alert the Base and set phase
> A of his brand-new, and detailed, Base warning system into

effect. We were more than delighted with this approach, knowing in our hearts that we were "off the hook" since this would cover us. The chance of a second tornado hitting the same spot within five days was less than 1 in 20,000,000. Far better we should take such odds rather than actually issue a tornado forecast and be laughed out of Uncle Sam's Air Force. We issued the General's heavy thunderstorm warning—what else?

Late morning turned to afternoon. A line of thunderstorms developed over western Oklahoma and began moving toward Tinker. General Borum was notified a second time.

The General spent ten minutes scanning the radar scope and commented on the rapid development and increasing intensity of the squall line. By 2:30 p.m. we determined the line was moving toward Tinker at 27 mph, which would place it over the base near 6:00 p.m.

E. J. and I glanced rather apprehensively at each other, sensing what was going to happen next. General Borum stood up, looked us in the eye, and asked the unsettling question, "Are you going to issue a tornado forecast?"

I knew E. J. would come up with a sensible, honest answer, and he did. "Well, Sir, it sure does look like the last one, doesn't it, Bob?"

I tried to think of a brilliant answer and found myself saying, "Yes, E. J., it is very similar to last week."

The General was not particularly impressed with this intelligence: "You two sound like a broken record. If you really believe this situation is very similar to the one last week, it seems logical to issue a tornado forecast."

We both made abortive efforts at crawling out of such a horrendous decision. We pointed out the infinitesimal possibility of a second tornado striking the same area within twenty years or more, let alone in five days. "Besides," we said, "no one has ever issued an operational tornado forecast."

In fact, Joe Audsley of the Weather Bureau had issued a tornado message in Sioux City the year before, but it was concealed for fear of disciplinary measures.

> "You are about to set a precedent," said General Fred S. Borum. With a sinking feeling in the pits of our stomachs, E. J. composed the historic message and I typed it up and passed it to Base Operations for dissemination. The time was 2:50 p.m. The General left, asking to be kept informed of significant developments. We discussed our suddenly impossible predicament. It seemed a hopeless situation; one where we could not win and the General couldn't lose. Base Personnel were carrying out his detailed Tornado Safety Plan, hangaring aircraft, removing loose objects, diverting incoming air traffic, and moving base personnel, including the control tower personnel, to places of relative safety. I could see it now, a sure "bust" and plenty of flack thereafter. I figured General Borum wasn't about to say, "I made them do it." More likely it would be, "Major Fawbush and Captain Miller thought it looked a great deal like the 20th—ask them." I wondered how I would manage as a civilian, perhaps as an elevator operator. It seemed improbable that anyone would employ, as a weather forecaster, an idiot who issued a tornado forecast for a precise location.

Every meteorologist, tasked with forecasting or warning of tornadoes, has grappled with similar thoughts.

> The squall line was fully developed by half past three and continued to move steadily toward Oklahoma City. There had been no reports of tornadoes nor any reports of hail and high winds, as yet. We were both very apprehensive and at

this point would settle gratefully for a loud thunderstorm with a brilliant lightning display and hopefully a wind gust to 30 or 40 mph with perhaps some small hail. General W. O. Senter, commander of the Air Weather Service [in a quirk of organization, the meteorologists at an Air Force base did not report to the base commanding officer, they reported to the head of the Air Weather Service], would perhaps be more merciful if we could just get a reasonably heavy thunderstorm. Shortly after 5:00 p.m. the squall line passed through Will Rogers Municipal Airport, but this time they not only didn't report a tornado, but infinitely worse, a light thunderstorm, wind gusts to 26 mph, and pea-size hail. That did it. I abandoned ship, leaving a grim Major Fawbush to go down with the vessel.

I drove directly home. E. J. and I both lived in Midwest City, just across the highway on the north side of the base. I related the events of the day to my wife, Beverly, who was reasonably sympathetic, and then sat down to aggravate my depression systematically. A little after six o'clock it began to thunder rather quietly and rain began. There was very little wind. It became quite dark, and over the Base, portions of the clouds seemed to be boiling while low cloud fragments darted hither and yon beneath the base of the thunderstorm. My view was quickly obscured by heavy rain and I stopped observing the storm.

During the evening, the radio broadcast . . . was interrupted for an urgent news bulletin. I was in another part of the house but caught the words destructive tornado and Tinker Field. "Good grief," I thought, "they're still talking about last week's tornado—but why break into the news?" I tried to call the weather station but the lines were dead.

I felt a strange unbelieving excitement rising, told my wife I was going to the station, and drove away. The Base was a shambles. Poles and power lines were down and debris was strewn everywhere. Emergency crews were busy trying to restore power, clear the streets, and in particular, to restore the main runway to operational status. I reached the station to find a jubilant E. J. who described the course of events after I had given up hope. At six o'clock thunder began at the Base as the squall line moved in from the southwest. E. J. and my friend the Sergeant were outside, observing the motion of the clouds. As the line approached the southwest corner of the field, two thunderstorms seemed to join and quickly took on a greenish black hue. They could observe a slow counterclockwise cloud rotation around the point at which the storms merged. Suddenly a large cone-shaped cloud bulged down, rotating counterclockwise at great speed. At the same time, they saw a wing from one of the moth-balled World War II B 29's float lazily upward toward the visible part of the funnel. A second or two later the wing disintegrated, the funnel shot to the ground, and the second large tornado in five days began its devastating journey across the Base, very close to the track of its predecessor.

It was all over in three or four minutes. It seemed much longer. The swirling funnels left $6 million in damage, $4 million less than the first storm, and significantly, there were no personal injuries. General Borum's Tornado Disaster Plan had been just as successful as the first operational tornado forecast. We became instant heroes, and in my case, the rest of my life would be intimately associated with tornadoes and severe thunderstorms. General Borum graciously refrained from mentioning the story behind the sensational forecast, and he convinced General Senter that we should be allowed to concentrate on further development of our forecast system. . . . The complexity and evolution of the pattern that instigated the sequence of events I have described boggles the mind. This first tornado forecast triggered a chain of events which led to the present day Severe Storms Forecast System and a vast national research program investigating these killer storms.

Well, it did look a lot like March 20. Even the General thought so.

*   *   *

Until his death, Robert Miller was very proud of his tornado forecasts, and with good reason. The Tinker tornado on the 25th appears to have been the first of a series of tornadoes in eastern Oklahoma and western Arkansas that killed thirteen people and injured forty-four. Had the base not been prepared by his forecast, it is possible that Tinker personnel might have been added to the number of casualties.

Yet at the same time Miller found himself conflicted—almost to the point of embarrassment—about them. He described something familiar to many forecasters when he realized his busted first tornado forecast might not have been a bust, after all, and said, "I felt a strange unbelieving excitement rising . . ."

It's natural to want to be proud of one's work. When a surgeon repairs a knee, when an air traffic controller guides an aircraft in distress to a safe landing, or when a wide receiver catches a touchdown pass, the emotion is celebratory, even jubilant. The airliner's captain keys the microphone and utters a terse "thanks" as she wipes the sweat off her brow. Coworkers or teammates offer pats on the back. The crowd cheers.

How would you feel if you were a meteorologist who had just forecast a tornado? Would you be happy if no tornado occurs, though you'll be bombarded by criticism from those listening to your forecast? Or do you hope for the tornado to make good on your prediction?

Miller wrote about this internal struggle in what he called a "plea for understanding":

The close-knit world of the tornado and severe thunder-storm forecaster often seems somewhat demented to those not knowledgeable in this discipline. The apparent derangement is based on our seemingly ghoulish expressions of joy and satisfaction displayed whenever we "verify" a tornado forecast . . . There is a fantastic feeling of accomplishment when a tornado forecast is successful. We are really nice people, but odd.

Among meteorologists in a forecast center or during a research storm chase, language inverts; "bad" becomes "good":

Meteorologist: "This looks like a terrific outbreak."

Translation: "There will be many violent tornadoes."

Meteorologist: "It was an awful day."

Translation: "I forecast tornadoes and none occurred."

Meteorologist (arriving back from a storm chase): "What a great day!"

Translation: "I saw an impressive tornado."

I have worked with and counseled meteorologists who were terribly upset with their own behavior around accurate tornado forecasts. In the heat of the moment, some "high-fived" each other, offering congratulations on the accuracy of the forecasts they'd made. When they subsequently learned of the storm's fatalities, their emotions instantly plunged from exhilaration to sadness and depression.

Meteorologists are extremely public-service-minded and want to prevent deaths. In situations where fatalities occur, they exhibit a form of survivor's guilt: "Maybe if I had just broadcast the warning one more time . . . If I had just emphasized *go to the basement* more . . ." In most of these situations, they did their job well and those who were injured either didn't get the warning or chose to ignore it.

Nevertheless, the "do you wish for a tornado after you have forecast it?" question is another unique aspect of meteorology. As Miller put it, "we are really nice people, but odd."

* * *

To get back to the story: The Air Force began immediately to capi-
talize on the new knowledge acquired from Miller to protect their
facilities, aircraft, and people from tornadoes. Fawbush and Miller
took the lead and became celebrities in the meteorological community. In 1951, the Air Force set up the first official tornado and severe
thunderstorm forecasting center.

The Weather Bureau's ban on civilian tornado forecasts and warnings continued, but it wasn't possible to keep a lid on the Tinker Air
Force Base tornado forecasts. Word about the Tinker forecasts got
out. Oklahoma City TV stations and Air Force meteorologists compared notes on a number of occasions. After the success at Tinker,
meteorologists started referring to tornadoes in oblique ways, saying,
for example, "these thunderstorms could be especially strong today."

Pressure began to build on the Weather Bureau and on local television stations to broadcast tornado forecasts and warnings. According
to Harry Volkman, the chief meteorologist at WKY-TV, the Oklahoma City NBC affiliate, the most intense pressure came from the
wives of the Tinker meteorologists who lived off base. "Don't we
deserve to know [if a tornado is coming]?" they asked. T. A. "Buddy"
Sugg, the general manager of WKY-TV had had enough. He told
Volkman that he wanted to put the Tinker forecasts on the air.

Using today's nomenclature, Fawbush and Miller created the first
tornado *watch*. A watch is issued hours in advance to alert people over a
relatively large geographic area that tornadoes are possible, and to give
them time to take preliminary precautions (e.g., move lawn furniture
and other outside objects that can blow around indoors) and to tune in
to radio or television broadcasts should thunderstorms approach.

But the Weather Bureau still refused to have anything to do with
Tinker's forecasts. Air Force regulations forbade giving the forecasts
to nongovernment agencies or companies. So Sugg and Volkman came
up with a plan: they dispatched reporter Frank McGee (later of NBC

News fame) to Tinker on March 25, 1952, to do a "news story" on their tornado forecasting methods. McGee sat next to Fawbush and Miller for hours and called in to Volkman periodically with updates. Sugg instructed Volkman to be professional should he have to air a tornado warning and "don't overemphasize it or scare people."

The call finally came from McGee: Tinker believed that tornadoes were going to develop in central Oklahoma as thunderstorms began forming. Volkman went on the air: "Tornadoes may develop in the WKY viewing area . . ." This first broadcast tornado warning was a success. According to Volkman, the tornadoes came.

In Washington, D.C., Weather Bureau chief Francis Reicheldefer hit the roof. It was bad enough that the military was making tornado forecasts; having civilians forecasting and warning of tornadoes offended the bureaucrat's sense of territory. The Bureau didn't want to make tornado forecasts, and it didn't want anybody else doing it, either. Volkman later described the flack as a "firestorm."

Reicheldefer flew to Oklahoma City to demand that WKY discontinue the warnings, even though he had no legal authority to do so. The station management cited the First Amendment.

A compromise of sorts was reached. Reicheldefer met with various officials in Oklahoma City, including the police and fire departments, but *without* Volkman, as the Weather Bureau of the 1950s did not look kindly on private sector meteorology. Everyone agreed that WKY would continue the warnings but that the Weather Bureau disclaimed any responsibility for them.

By 1956, WKY added a converted aviation radar to its arsenal of weather tools and for several decades continued to issue tornado warnings independent of the Weather Bureau.

* * *

The Weather Bureau found itself isolated. The military was issuing tornado forecasts. Now the media was issuing tornado warnings.

The Weather Bureau reluctantly started exploring the field of tornado forecasts and warnings.

In March 1952, the Bureau analysis center in Washington established a Severe Weather Unit (SWU). On March 17, it issued its first civilian tornado forecast (in today's nomenclature, a *watch*). Four days later, it issued its first successful (that is to say, accurate) forecast. But these forecasts were primarily intended as *internal* guidance material to the Weather Bureau field offices; the SWU was not even in business twenty-four hours a day. Perhaps the Weather Bureau thought it could stave off its critics with this half-hearted attempt to forecast tornadoes.

But the events of 1953 would convince even the Weather Bureau that the clock on its delaying tactics had run out.

CHAPTER FOUR

THE GOVERNMENT GETS IN GEAR

THE TORNADOES KEPT COMING, AND THEY WEREN'T confined to Tornado Alley. The 1952 fledgling tornado forecasts from the Weather Bureau were having some modest success in raising public awareness, but that was about all.

There was still reluctance by the Weather Bureau to share its full output with the public. So the tornado forecasts, when issued, went through a complex and slow-paced review process at its field offices and only sporadically reached the public.

That changed after three tragic days in 1953.

* * *

May 11. A thunderstorm complex produced a tornado in Tom Green County, Texas, that moved into the town of San Angelo, killing thirteen people. Farther east, two hours later, a massive tornado developed near Lorena and moved directly into and across downtown Waco.

This F5 tornado completely devastated the center of the city. It killed 114 people and injured 600 more.

Four weeks later, a major storm system moved across the Upper Midwest. Tornadoes developed in Iowa and Nebraska on June 7. On June 8, a series of tornadoes moved through Michigan and Ohio. The worst storm struck the city of Flint at about 8:30 in the evening, killing 115 people and injuring 844. An additional twenty-three people were killed by tornadoes at other locations in the two states.

The broadscale weather system continued to move east. Tornadoes developed in Massachusetts and New Hampshire the next day. An F4 struck the town of Worcester, Massachusetts, killing 94 people and injuring 1,288. At one point, the Worcester tornado was a full mile wide.

With hundreds dead and thousands of people injured, the public and press grew tired of excuses and rationalizations. Even the set-in-its-ways Weather Bureau realized it was time to set up a genuine tornado forecasting program and forced its Severe Weather Unit (SWU) in Washington to transition immediately from a quasi experiment to an operational unit. Eight days after the Worcester tornado, SWU issued its first truly public forecasts, but they came at a price: the staff of five was pushed to the limit by the swarms of tornadoes. They handled the Texas and Midwest tornadoes well (especially considering the state of the art at the time), but the New England tornadoes were a surprise. Of the five-man staff, one forecaster resigned almost immediately, and a second quit prior to the end of the year. The pressure to get better, and get better fast, kept increasing.

In 1954, the SWU moved from Washington to Kansas City, due primarily to media influence, the perception being that Kansas City was closer to the tornado action. Given the 1953 communications systems, that perception was close to reality. Kansas City was a major switching center for teletype circuits carrying meteorological information around the country, and so could receive critical data faster and disseminate forecasts more quickly than could other locations.

SWU had a new director, Don House, and a new name: the Severe Local Storms Unit became known as SELS and issued the tornado forecasts (today called tornado watches). These and other meteorologists based in Kansas City published important new science in the 1950s. The accuracy of tornado forecasts slowly started to improve.

In spite of public praise for SELS and the concentration of resources and expertise in Kansas City, the reticence of the Weather Bureau continued.

Steve Corfidi, a former SELS forecaster, explains:

> During this period, a typical SELS tornado forecast would read as follows: "possibility of an isolated tornado along and thirty miles either side of a line from Amarillo, TX, to 20 miles north of Gage, OK, from 5:15 to 9:00 PM." Such a forecast would have first been telephoned to the district offices(s) involved. If it was agreed that a public forecast of tornadoes was indeed prudent, the district forecaster would notify the local Weather Bureau offices under his jurisdiction, in addition to the media. If the proposed forecast affected only one district office, that office had final say as to whether or not tornadoes would be mentioned in the public forecast. If, on the other hand, a proposed tornado forecast involved more than one district office, SELS made the final decision. It was not until 1958 that SELS assumed total authority for public tornado and severe thunderstorm forecasts.

In other words, in spite of the increasing expertise at SELS, all it took was one old-school field forecaster to block a forecast from reaching the public! The pressure of these frustrating procedures continued to take its toll. By 1955, only one of the original forecasters, Galway, was still working with the program. And in that year, the failure of that cumbersome system may have cost more than 100 lives in a single evening.

CHAPTER FIVE

THE "TOWN THAT DIED
IN ITS SLEEP"

ON MAY 25, 1955, DON BURGESS SAW HIS FIRST tornado. A spectacular thunderstorm passed just west of the city of Stillwater, Oklahoma. To this day, Burgess remembers the lightning. He was standing outside one of the student housing complexes at Oklahoma State University watching the thunderstorm around eight o'clock in the evening with his parents and a number of other adults. In the waning light, lightning illuminated a tornado in the northwest that had just dropped from the thunderstorm.

Even in his earliest years in Okmulgee, Oklahoma, Burgess was interested in weather: when a storm approached, he and his father would climb onto the roof of their home to watch the sky. In 1955, his family moved to Stillwater and Burgess would spend his afternoons playing with his friends; when it was time for the weather report to come on television, however, he ran inside to watch.

May 25 had been a day of variable weather, with off-and-on showers and thunderstorms in central and northern Oklahoma. A small funnel cloud had been reported north of the town of Pawnee around 4:20 p.m. It went back up into the clouds without causing any problems.

At 4:36 p.m., the Weather Bureau office in Wichita transmitted a tornado forecast from the SELS unit in Kansas City. Wichita ended its local loop teletype bulletin with "immediate broadcast is desired. Please and thank you." Remember, this was still the era of tornado forecasts having to go through multiple gatekeepers to reach the public. This forecast, called a "warning" in 1955, was essentially identical to today's tornado watch.

Unlike the more-or-less ordinary showers and thunderstorms seen earlier in the day, a supercell thunderstorm formed over Oklahoma City shortly after six that evening. It passed by Stillwater around eight in the evening, moving toward the north-northeast. This was the storm that spawned the tornado seen by young Burgess, the first of four it would eventually create.

Burgess' memory about the lightning was accurate. Scientific papers would later discuss the amazing lightning produced by this supercell, including one by H. L. Jones published in *Weatherwise* magazine that I read and reread as a teenager. Cloud-to-ground lightning rates were as high as 25 strokes per second, with arcs of lightning reported from Payne and Kay Counties. The research on this storm pointed meteorologists on a quest, one that continues today, to understand the relationship between lightning and tornadoes.

At about 9:00 p.m., this unusual thunderstorm dropped a second tornado near the small town of Tonkawa, Oklahoma, destroying a few homes and outbuildings, and then dissipating. The FAA weather observer at the Ponca City Airport, northeast of the path of the second tornado, reported east-southeast winds (inflow into the supercell, bringing warm humid air, the fuel needed to sustain the thunderstorm) gusting to 35 miles per hour and "continuous" lightning. At this point the storm, still moving just a little east of due north, crossed to the east of today's Interstate 35, which put it *outside* of the tornado watch and into an area that had no idea a tornado might be coming.

At 9:25 p.m., a third, stronger tornado developed two miles northeast of Tonkawa and moved through the east side of Blackwell, Oklahoma. As the storm approached Blackwell, there was a period of rain accompanied by hailstones two inches in diameter. Instead of the air being cooled by the rain, the feel of the air was described as "hot."

The air was deadly still, but in the distance the "sound of forty freight trains," according to a Mr. Nave (first name unknown), rumbled in from the south and grew louder as the seconds ticked by. The tornado was illuminated by the frequent lightning, which sent the few people viewing the storm scurrying for shelter. The tornado hit without warning.

The F5 tornado destroyed 400 homes, many of which were swept away down to their slabs. Another 500 homes were damaged, with sixty businesses damaged or destroyed. Twenty people were killed, and 280 more were injured. The tornado continued north-northeast until it got to the vicinity of the Kansas border, where it lifted.

The police in Ponca City called the local airport weather station to tell the weather observer that a tornado had struck Blackwell at 9:30 p.m. The weather observer dutifully recorded this information and sent it out on the Service A teletype. And then, unlike the actions of the Pampa weather observer eight years before, *this* weather observer picked up the phone and called the Weather Bureau office in Wichita. The Wichita office had warning responsibility for southern Kansas, where the storm was headed.

At 9:50 p.m., the SELS unit issued a second tornado forecast for farther east, but Blackwell had already been hit. Worse, the Weather Bureau's bureaucratic hoops (the SELS forecast had to be first cleared by the district office in Denver before it could be sent to the local offices in Kansas) delayed the receipt of the forecast at the Wichita office until 10:08 p.m. This delay was critical. In the central time zone, the late news starts at 10:00 p.m.

Local Wichita television had already broadcast that "all advisories had been lifted." Because news personnel were in the studio, the 10:08

bulletin extending the tornado watch, at least at some stations, sat on the teleprinter unseen. Many people throughout the area turned in for the night once they had received the assurance that no advisories were in effect.

As the Blackwell tornado dissipated, the fourth tornado from this supercell thunderstorm touched down west of Arkansas City and continued on the same northward trajectory. The time was approximately 10:15 p.m.

On a farm outside the tiny town of Oxford, Kansas, Ruth King was feeling unsettled. She had been keeping track of the weather all evening after hearing a forecast of "severe thunder and electrical storms" on the radio between 5:30 and 6:00 p.m. After assuring herself there were no tornado warnings, she put her children to bed at nine o'clock, and she herself turned in about 9:20 p.m. At the time, it was raining with "lots of close lightning." At 9:50 p.m., she got out of bed to close the window because a strong wind started blowing from the east and golfball-sized hail started hitting the house. In her own words, she was "extremely nervous" but tried to calm herself after going outside and seeing that things didn't look "too bad."

* * *

From the Wichita Weather Bureau office, located on the southwest side of the city at what is now called Mid-Continent Airport, the WSR-3 radar was scanning the skies of south-central Kansas. The radar, acquired in 1946 from the Navy, was modified before being deployed by the Weather Bureau. Wichita was one of a handful of weather offices with that early-model radar. The six-foot-diameter contact lens–shaped antenna made continuous 360-degree sweeps searching for storms across the area.

Meteorologist Ellis Pike was on duty at the Wichita Weather Bureau that evening. Years later, Pike told me that the storm was very strong when it first appeared on radar in the area of northern

Oklahoma, and they were carefully watching it as it approached the Kansas border.

About 9:25 p.m., it began sprinkling at Mid-Continent Airport. By 10:04 p.m., just after the local news went on the air, heavy rain was falling at Mid-Continent. A few minutes later, the Weather Bureau noted the call pertaining to the Blackwell tornado and that "they requested aid from all towns."

There was probably a small sense of relief inside the Weather Bureau because it looked like southern Kansas would dodge the bullet that had struck Blackwell. Pike said the radar echo coming out of Oklahoma was weakening and shrinking by the time they heard about the Blackwell storm.

By now, heavier rain whipped by the wind was striking the windows of the small Weather Bureau airport office. Unknown to Pike, a thunderstorm was also in progress at McConnell Air Force Base (AFB) on the southeast side of Wichita. It was producing even heavier rain than the storm over Mid-Continent.

The WSR-3 radar painted greenish white "blobs" of weather on a cathode ray tube. The radar had to be operated in a completely darkened room, since radar echoes faded very quickly behind the radar "sweep" as it rotated around the screen in sync with the rotation of the radar antenna. Eyestrain was an occupational hazard.

By 10:10 to 10:15 p.m., the storms over Mid-Continent Airport and McConnell AFB, and the Oklahoma supercell that had moved near the town of Oxford were starting to configure themselves into a straight line. While the WSR-3 was designed to penetrate rain, it was not powerful enough to penetrate a solid 40-mile-thick area of intense thunderstorms. And the weather was going from bad to worse at McConnell AFB, with torrential rain and wind gusts up to 45 miles per hour by 10:17 p.m.

A weather radar operates by sending a brief "pulse" of focused microwave energy from the antenna (think of a flashbulb going off) to search for precipitation—raindrops, hailstones, snowflakes. When the microwave pulse encounters precipitation particles, it "scatters," with part of the energy reflecting back to the radar. Many radar antennae are shaped like contact lenses to focus the radar's pulse. The greater the focus, the more the details in the storm can be depicted.

The radar works in a pattern of *pulse . . . listen* (i.e., listen for the reflected energy to return to the radar) *. . . pulse . . . listen*, with the cycle repeating 200 times or more per second.

When the supercell was near Blackwell, there was little or no precipitation between the supercell and Wichita, and the supercell appeared on the radar as a monster storm; the bright white blob was as white as the radar could paint it. As the supercell traveled north and aligned with the two smaller thunderstorms at Mid-Continent and McConnell, the mass of dense rain and hail between Mid-Continent and the supercell increased. Thus, in the critical minutes between 10:00 and 10:20 p.m., less and less reflected energy made it back to the radar's antenna.

When a radar signal is absorbed by precipitation, the result is called *signal attenuation* (or just attenuation). As the reflected energy from the supercell received by the radar lessened, the supercell appeared to be getting smaller and smaller, weaker and weaker.

The incorrect conclusion at the Wichita Weather Bureau was that they didn't have to worry about the supercell any longer. The shrinking radar echo (now near Oxford) observed by Pike was an illusion. The reality over far eastern Sumner County was that the supercell thunderstorm was stronger than ever.

Wichita radio station KWBB had seen the tornado forecast on its teletype and broadcast it around 10:15 p.m. However, this was AM radio in 1955, and KWBB wasn't a powerful clear-channel station. (There are three types of AM radio stations: clear-channel stations, allowed to broadcast with full power at night; those that reduce power at night; and those that sign off the air at sunset.) With no clear-channel stations in Wichita, it would have been impossible for anyone in rural Sumner or Cowley Counties to have heard the KWBB severe weather forecast. And without a forecast, no one would have thought to be outside looking at the sky.

* * *

Back at the King farm, Ruth King noted the east wall of their house shaking around 10:20 p.m. Ruth ran to gather up her five children, but as she did so, her house began "flying into pieces." King and her children—along with what was left of the house—began flying in every direction. As Ruth flew through the air, she screamed for "God to have mercy" but found the wind "tore the words out of my mouth." When she came to rest she kept saying, over and over, "I can't believe it, I can't believe it, I can't believe it. . . ." Pieces of their furniture were found more than three miles away.

The memory of the bodies of the five dead King children scattered over the farm field would haunt their would-be rescuers for decades.

* * *

This fourth tornado from the immense supercell that had formed over Oklahoma City more than three hours earlier slammed into the little town of Udall, population 500, without any warning, at 10:35 p.m. another seventy-seven people died, most of them killed by flying debris. Several, who dashed into their basements when they realized a tornado was destroying the town, drowned when the town's water tower toppled and its water flooded their underground shelters.

This tornado was a strong F5. "Udall died," a prize-winning headline later proclaimed, "in its sleep." There was only one habitable building left in the city.

Devastation caused by an F5 tornado in Udall, Kansas, May 25, 1955. Courtesy
of the Udall Tornado Museum.

For more than an hour, no one outside of Udall knew what had happened. Finally, around 11:30 p.m., the Wellington police department
in Sumner County received a call stating that Udall had been hit. It
took another thirty minutes before calls were made to Wichita and
the surrounding communities summoning aid. Wichita radio news
reporter Henry Harvey made it to Udall sometime around one in
the morning, and he described the experience as "the worst night of
my life." The crying children, the "continuing moans of the yet to be
discovered," and the shouts of rescuers moving wreckage—(excitedly)
Here are some more! (deflated pause) *But I think they are all dead*—
"drummed through his head" long after he left Udall to report on the
disaster.

By the end of the night, tornadoes number three and four from
the Blackwell-Udall supercell had killed 102 people and left 1,300
injured.

Chevy truck wrapped around tree stripped of bark in Udall. The driver's body
was found 0.25 miles outside of town. Courtesy of the Udall Tornado Museum.

* * *

As tornado after tornado in the 1950s produced triple-digit death
tolls, it was becoming obvious that converted World War II aircraft
radar was not sufficient for dealing with the tornado problem.

In fairness, the Blackwell-Udall supercell was not the first time
that radar data had been misinterpreted. At Pearl Harbor, attacking

Japanese planes were detected on radar as they were heading south across the Island of Oahu, but the supervisory radar operator thought the planes were a flight of U.S. aircraft headed to Pearl Harbor from the mainland and failed to sound the warning. So the naval fleet and sailors stationed at Pearl Harbor were sitting ducks, just like the people in the path of these tornadoes.

For a number of reasons, radar was (and in some ways still is) difficult to interpret. Here's an analogy: How many times has a doctor described an x-ray and, looking at the white-and-black film hanging on the light board, you didn't have the slightest idea what she was talking about? Then, after a hard-to-follow explanation about "shadows" and "thoracic and pericardial effusion," the doctor says, "We aren't sure, so we must do some exploratory surgery or perhaps run an MRI."

Medical x-rays, like weather radar, can be ambiguous. But, unlike doctors, meteorologists don't have the luxury of running more tests. Meteorologists have to make split-second decisions when tornadoes, flash floods, and other life-threatening conditions present themselves.

As Don Whitman, assigned to the Wichita office six months after Udall, explained, "we simply did not understand radar attenuation at the time," which is why they were lulled into a false sense of security as the tornado bore down on Udall. Even if attenuation had been understood, it would have been difficult to issue an alert for Udall; attenuation distorts the way the storm is presented on the radar, so the Wichita meteorologists would have been unable to determine the exact location of the tornado. Plus, it was 1955 and the Weather Bureau's ban on issuing what we now call a tornado warning was still in effect.

Frustratingly, Tinker AFB's radar, 130 miles away, *was* tracking the Blackwell-Udall supercell, as indicated by the accompanying radar tracings. Even with Tinker's much greater distance from the storm, its radar was able to see the supercell because the sky was otherwise clear (no intervening storms) between Tinker and the supercell.

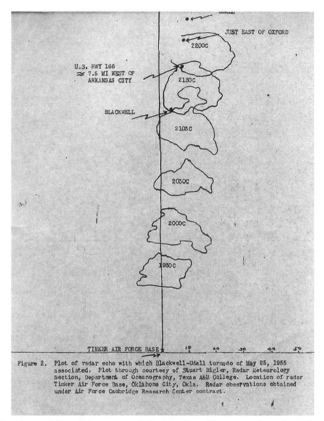

Figure 2. Plot of radar echo with which Blackwell-Udall tornado of May 25, 1955 associated. Plot through courtesy of Stuart Bigler, Radar Meteorology Section, Department of Oceanography, Texas A&M College. Location of radar Tinker Air Force Base, Oklahoma City, Okla. Radar observations obtained under Air Force Cambridge Research Center contract.

Tracing the position of the supercell that produced four tornadoes, including those at Blackwell and Udall, at half-hour intervals. Udall is at the very top of the image. The tracings were made at Tinker Air Force Base's weather station. Courtesy of TornadoChaser.com.

Because attenuation was not understood in 1955, it would never have occurred to the Tinker meteorologists to call Wichita—after all, they likely assumed Wichita could see the violent thunderstorm just fine, since Wichita was so much closer. But even if the Wichita meteorologists had understood that attenuation was blotting out the Udall thunderstorm, the technology did not exist to transmit the Tinker radar images outside of the Tinker weather office. And at the time of the Udall tornado, the Weather Bureau still regarded the Air Force with suspicion.

A better radar, and a better way, had to be found.

* * *

The British invented radar (its name is an acronym for RAdio Detecting And Ranging) in the late 1930s. At first, weather appearing on radar was considered a nuisance: it obscured the detection of air-craft. As World War II progressed, however, people started realizing that radar could be a storm-warning tool.

How are storms located by radar? The process seems simple: We know the direction the radar antenna is pointing. If the antenna happens to be pointing due east, anything detected by the radar at that time is along the 90-degree corridor. We know the distance from the radar because we know the speed of light, so we just time the interval (which is counted in millionths of a second) from when the pulse leaves the radar to the time it returns, and then convert time to distance from the radar.

After several years of development, in 1959 the Weather Bureau deployed the first radar specifically designed to monitor weather. The Weather Surveillance Radar-57 (WSR-57) was a significant step forward. It was more powerful than earlier radars (improved capacity to penetrate storms), was able to measure the storms in three dimensions, and had a camera to record what the radar was displaying: radar images could be analyzed after the fact so that signatures common to storms could be shared and researched.

The first WSR-57 was deployed in Miami, Florida, with Kansas City (partly because of the Ruskin Heights tornado) next in line. The plan was for the Kansas City meteorologists (including meteorologists with local warning responsibility and SWU meteorologists) to be able to monitor a vast area of the Midwest: Oklahoma City, Tulsa, Topeka, Wichita, Omaha, Des Moines, St. Louis, and Springfield. So it was decided to put the radar as high as possible, above any obstructions, to take advantage of its greatly improved range. The radar was

mounted high above downtown Kansas City, on top of the federal
building at 901 Walnut Street.

WSR-57 radar in downtown Kansas City. The white "ball" is a protective dome
over the radar. The radar's height created so much ground clutter, Kansas City's
weather was poorly visible on the radar. Courtesy National Weather Service.

Two years after Ruskin Heights, Kansas City had the best weather
radar in the world. There was just one problem: the radar could not
see Ruskin Heights!

It is intuitive to think that the higher up a radar is located, the
less ground clutter (nonweather echoes such as trees, hills, and buildings) the radar will display. This is intuitive, but wrong. It turns out
that ground clutter *increases* when the radar is located high above the
ground. So while the Kansas City radar could see Topeka, St. Joseph,
Sedalia, and other towns in the region, it could not effectively see

Kansas City and its suburbs. Eventually, a secondary radar was located in Topeka, fifty miles to the west of Kansas City, and the status of the Topeka weather office was upgraded, so that Kansas City would have effective radar coverage.

It was even worse at Missoula, Montana: That radar, on top of a mountain, had ground clutter for a radius of nearly 110 miles in all directions.

Even with ground clutter problems in Kansas City, Missoula, and a few other locations, the WSR-57 was a major step forward. Miami proved to be a fortuitous early location, as Hurricane Donna was effectively tracked by the radar in 1960, resulting in improved warnings for south Florida. Hurricane Carla was well tracked in 1961 by the Galveston WSR-57. There is no doubt this improved tracking of storms led to better warnings and saved lives.

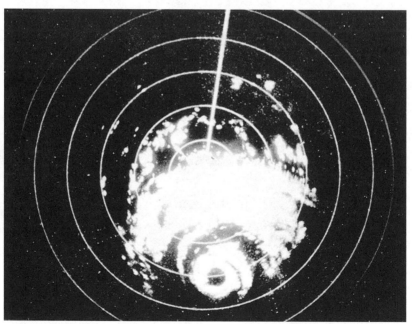

The eye of Hurricane Carla as depicted on the new WSR-57 radar. Note the lack of mapping and the poor contrast. Nevertheless, the WSR-57 was a major step forward.

* * *

After the example set by Joe Audsley and the success of his warnings during the Ruskin Heights tornado, other field offices began defying the Weather Bureau's mandate and started issuing tornado warnings. While it was still a difficult process, accurate tornado warnings were starting to become a reality as meteorology moved from the 1950s into the 1960s.

Perhaps the first complete success was the Topeka, Kansas, tornado of June 8, 1966. A tornado watch had been out for northeast Kansas, including Topeka, for hours and had been widely broadcast. Just about everyone knew that tornadoes were likely in the area that afternoon and evening.

Then a tornado warning was issued for Topeka. The TV news anchor, Bill Kurtis, told his audience, "For God's sake, please take cover!" Warning sirens wailed, the people of Topeka heeded the Weather Bureau's warning, and the tornado cut the predicted path across the entire city, including the downtown area where it damaged the State Capitol.

In many ways, the Topeka and Waco tornadoes were very much alike: F5 intensity, similar populations, downtown areas devastated, and so on. In Waco, the death toll was 114. In Topeka, the death toll was just 16. *This* was real progress.

The Weather Bureau basked in well-deserved praise after the Topeka storm. Editorials were written. Grateful letters from survivors were received. The new radars were having a positive effect.

In Oklahoma, the Weather Bureau had founded the National Severe Storms Laboratory and was testing new radars and instrument arrays. The nation was safer from tornadoes.

Still, a significant number of tornadoes were invisible to conventional radar. The solution to that problem would come from young Don Burgess, who had seen the first of the tornadoes from the Blackwell-Udall supercell.

* * *

The year 1957 exhibited two high-profile tornadoes. Ruskin Heights has already been discussed. It was well forecast and, because the tornado was both on the ground a long time and under the Kansas City field office's radar umbrella, was well warned, considering the policies and state of the art at the time.

Six weeks before Ruskin, Dallas was hit by the only major tornado to occur within its city limits. On April 2, a slow-moving tornado with a high cloud base and excellent lightning was visible from much of the burgeoning metroplex. It killed ten people.

Undoubtedly, the high-profile Dallas and Ruskin Heights tornadoes were part of the impetus to cut out the middleman. For the following spring's tornado season (1958), the Weather Bureau decided to deliver the SELS forecasts directly to the public via the media.

The tornado watch program continued to mature and improve. Rechristened the National Severe Storms Forecast Center (NSSFC), the leadership changed in 1965 when Don House retired from the director's position and was replaced by Allen Pearson, who raised the visibility of the watch and warning program through skillful public relations and dealings with the media.

Pearson also started a Kansas City amateur weather club that met on Saturdays to encourage me and other area teenagers to learn more about weather and storms.

While NSSFC was in Kansas City, I was a frequent visitor and privileged to meet Robert Miller. The Air Force, for a time, colocated its severe storm forecast operation with that of the National Weather Service (the name changed from Weather Bureau in 1970) in downtown Kansas City. Colonel Miller was always gracious to this enthusiastic teenager, answering all of my questions, and he even gave me a copy of his famous Air Weather Service tornado forecasting manual.

Along with all other severe weather forecasters, I use the basic principles in that document even today.

Miller's work lives on every time a tornado watch is issued. The group he and Ernie Fawbush started at Tinker Air Force Base moved from Kansas City to Omaha and is now called the Air Force Weather Agency. It has a broad range of weather-warning responsibilities, including issuing backup tornado and severe thunderstorm watches for the public if there is a communications interruption to the National Weather Service Storm Prediction Center in Norman, which issues watches for the public.

Miller and Fawbush jointly received the American Meteorological Society's Clarence Leroy Meisinger Award. Miller was recognized for his lifetime achievements by the American Meteorological Society, which elevated him to the level of Fellow of the Society. Robert C. Miller passed away in 1998.

CHAPTER SIX

THE PAUL REVERE OF GRANDVIEW JUNIOR HIGH

THE RUSKIN HEIGHTS TORNADO WAS NOT MY family's first experience with extreme weather.

On July 13, 1951, a record flood of the Kansas and Missouri Rivers wiped out Ray Smith Studebaker, my grandfather's car dealership (where my father worked), located near downtown Kansas City. As the *Kansas City Star* put it, "the 1951 flood was Kansas City's biblical disaster." In some areas, more rain fell from May 1 to July 13 than normally fell in an entire year.

The 1951 flood was the costliest disaster to date in the history of the United States. It caused $3 billion of damage (in today's dollars) in Kansas City alone. An additional $3 billion of damage occurred in the surrounding region. Ray Smith Studebaker had no flood insurance because none was available in 1951. While the people of the Greater Kansas City area rallied heroically to help those affected, there was nothing like the federal disaster assistance available today. In those days, *wiped out* truly meant "wiped out."

While Dad was selling Studebakers, Mom worked for a printing company nearby. She was pregnant with me as they dug the dealership building and cars out from the muck once the waters receded. They had to borrow money from a bank in order to reopen. After careful consideration, my grandfather and my dad decided to move from the city proper to the suburbs and build the Ray Smith Ford dealership in Raytown, Missouri.

We were a typical late-1950s suburban family. I had two brothers (two sisters came along later). After the dealership moved, Mom stayed home and raised us with great love. Dad was vice president of Ray Smith Ford and, like many entrepreneurs, probably worked too many hours. We were a very happy family.

My neighborhood was perfect for a five-year-old boy. At the end of our street was an undeveloped area known as the Dead End. It contained a small stream (the storm water runoff from our neighborhood) in which to hunt crawdads and tadpoles, trees to climb, and the Kansas City Southern railroad track, all of which made it a boy's paradise. Because there was no street crossing in the area, the trains usually didn't blow their horns in our vicinity. We knew a train was approaching because you could feel a vibration in your shoes before the train came into sight. We would go running toward the railroad embankment and watch a Kansas City Southern freight or, better still, a beautifully painted *Southern Belle* passenger train go by.

Watching those trains got me interested in railroading. The interest grew when my uncle Everett Mealman took me to watch trains in downtown Kansas City on Sunday nights after an authentic Italian dinner at my Grandmother Lembo's. My love of railroading has stayed with me throughout my life, and it would eventually become an important part of my career.

Just a few weeks before that fateful Ruskin Heights evening in May 1957, I toured my kindergarten-to-be, Tower School, so named because it was almost directly underneath the Ruskin Heights water tower. I was looking forward to starting school when the Ruskin tornado damaged the school building. Amazingly, Tower School was repaired in time for kindergarten to start in September.

Soon after I learned to read I began going to the rebuilt library to read weather books. Grandmother Lembo gave me my first weather book, *Weather: A Golden Nature Guide*. I was getting a head start on my future career.

I learned that you could send away for Weather Bureau tornado brochures and other literature about weather from the Government Printing Office in Washington, D.C. They were typically about ten cents each, and I sent away for every brochure I could find.

My Christmas present from my grandmother and grandfather Smith the next year was a toy printing press. I thought it was a great gift: I could print weather forecasts! So after learning the intricate process of making it work, I produced official-looking—at least to my eye—weather forecasts. Being a child of generations of entrepreneurs, I easily took the next step. Now that I had a printing press and could print forecasts, what could be more natural than going door-to-door selling them?

In the 1950s, door-to-door sales were very common, so selling door-to-door was not as crazy as it might seem. What was crazy was trying to sell weather forecasts. Nevertheless, as soon as the ink dried on my freshly printed forecasts, I set out. How much did my forecasts cost? 5¢ each.

I went next door and received my first setback: Mrs. Furry's reaction was, "Don't you know that weather forecasts are free?" Undiscouraged, I kept plying my newfound trade of door-to-door weather salesman. I covered the entire neighborhood that day. I seem to recall selling only one forecast.

In 1965, my parents moved us to far south Kansas City, and I transferred to Grandview Junior High School. Over time, I met others who had interests in weather or aviation. I learned that the Federal

Aviation Administration (FAA) continuously broadcast the airport weather codes over a long-wave radio in broadcasts intended for pilots; you just needed a special receiver to hear them.

I saved my lawn-mowing money and purchased a receiver on sale for $68 (a tremendous sum for a teenager in the late 1960s). Through my library-based self-studies, I already knew how to plot the FAA reports in meteorological shorthand. I learned how to analyze the weather maps and make forecasts by reading other weather books.

Each morning I hand-plotted a weather map based on the airport weather observations broadcast by my special radio. Through years of practice, I could plot the temperature, dew point, wind speed, wind direction, pressure, precipitation, visibility, and miscellaneous remarks as fast as they could be read.

One April day in 1966, my personal weather map—plotted before I left for school—was looking especially ominous. I told my teachers and the principal that they should pay attention to the weather. They raised their eyebrows and dismissed me.

As the day progressed, the cumulus clouds billowed higher and higher into the atmosphere. The air was becoming warmer and more humid. Then we had a brief rain shower late in the day that cranked up the humidity to oppressive levels. The air began to have the tornado "feel" known to residents of the Midwest. I told my final-hour math teacher, Miss Chapin, that she and the other members of the faculty should really pay attention to the weather as the sky started to darken to the west. She told me to quit worrying and that I should pay attention to the lesson.

The western sky was leaden with a bluish tint when school let out a few minutes later. The oppressive tornado feel was off the charts. Though I typically rode the school bus, one of the eighth-grade teachers offered to take me directly home.

We got about two blocks west of the school when we reached a clearing in the buildings that lined Main Street in downtown Grandview. We had a nearly unlimited view toward the west. I was in the passenger seat, looking down, making an entry into my weather journal when Mr. Belauh said, "Well, look, there's a funnel cloud now." I looked up and saw a tornado to the northwest. It was blue-gray, starkly highlighted by partial sunshine to the west.

I gasped. "That's a tornado!"

"Are you sure?" Mr. Belauh asked.

"Yes!"

"We'd better get back to school to get the students inside," he said, making an abrupt U-turn with his car. In those days, the school bus came about forty minutes after school to allow time for extracurricular activities, and a number of students were milling around outside Grandview Junior High waiting for their rides. Just as we were getting out of the car back at the school, the tornado sirens came on.

Rain and then hailstones started falling, harder and harder. Everyone ran back into the building. After more than a half hour of our standing in the hallways, the all clear sounded in Grandview.

About ten miles to the northwest, in the Kansas City suburb of Overland Park, teachers and students at Katherine Carpenter Elementary School were emerging from shelter. The tornado had just missed the school. The teachers, believing there was dangerous weather potential at the time for school dismissal, had held their students and put them in shelter.

While there was no tornado warning yet in effect, the combination of the tornado feel to the air, the approaching thunderstorm, and a tornado watch caused the teachers to take action. Had the teachers dismissed school normally, the students would have been walking home while the neighborhood was destroyed around them. Many students owed their lives to those quick-thinking and decisive teachers.

Safe in Grandview, I found the experience exhilarating. I had forecast that tornado—hours in advance! Like Paul Revere before me, I had sounded the alert. I told people all day a tornado would occur and it happened just as my forecast said it would.

After that, people began taking my weather forecasts seriously.

My homeroom teacher had a friend, Norman Prosser, who worked for the Weather Bureau. A few weeks after school had let out for the year, he arranged to take me downtown to the new federal building to one of the hallowed halls of meteorology, the National Severe Storms Forecast Center (NSSFC). Mr. Prosser gave me the grand tour. I got to see the people who made the tornado watches. I got to see the radar room. And while we were in the Public Service Unit (PSU)—the unit that broadcast the local Kansas City forecasts and storm warnings— Mr. Prosser introduced me to the PSU head Don Whitman, who was to become my mentor and one of my best friends.

Mr. Prosser knew the Bureau needed a storm spotter in the far south part of Kansas City; he realized I was knowledgeable, so they ran it through their system for making a storm spotter official, and a few days later a U.S. government eight-by-ten envelope came in the mail with the unlisted phone number for the radar room!

I became Spotter 185/13. Spotters were designated by their direction, in this case 185 compass degrees (or slightly west of due south); and distance, 13 nautical miles from the Kansas City radar. By using that system, the radar operator could quickly correlate what the spotter was seeing with what was displayed on the radar.

More than a few times, I was out in the rain trying to spot tornadoes and call in reports of hail or high winds. I saw a funnel cloud once and reported it. Because there was also a hook echo, the radar signature of a tornado, the Weather Bureau issued a tornado warning and the sirens went off.

The next spotter to the northeast, a policeman stationed in Kansas City's Minor Park, also saw the funnel as it moved toward him. A few moments later, it went back up into the clouds without ever touching down.

A local newspaper, *The Squire*, ran a story about the high school student who warned Kansas City. My parents were proud, but they were torn between their pride and wanting their son in out of the rain.

CHAPTER SEVEN

THE END OF THE BEGINNING

THE TOPEKA TORNADO OF 1966 OCCURRED THE summer before I started ninth grade, the first year of high school in the parochial system. On January 24, 1967, over the noon hour, we were in the windowless Rockhurst High gym for Mass. We could hear thunder, heavy rain, and considerable wind. Hail started falling with the sound reverberating under the roof. I had a bad feeling, but the storm soon passed. As we exited the gym, I could see sunlight mixed with the thunderclouds through a glassed area connecting the gym to the school. I didn't know until I went home for the day that the storm had caused minor damage near Rockhurst.

Around the time we were leaving the gym, the same thunderstorm spawned a tornado that slammed into the Orrick, Missouri, high school. The gym collapsed with damage to other parts of the school. Two students were killed and many more were injured. As far as we knew, there was no warning of that storm. Watching the news reports at home that evening, I realized that if the thunderstorm had intensified a dozen miles farther west, it could have been our gym that was destroyed with the entire student body inside.

I transferred back to Grandview my junior year and started a teenage weather club called the Metropolitan Weather Service. We had about a dozen members and all had weather instruments and took readings twice a day. We broadcast our forecasts and weather observations on KMBZ Radio that, at the time, was the highest-rated radio station in Kansas City. I have to give the staff of KMBZ a great deal of credit for allowing local teenagers access to their very valuable airtime.

At Grandview High School I met Kathleen Rector. Our first date was not auspicious. I thought it had gone well; she thought I talked about nothing but weather. I saw her several more times because her mother was the only person with a complete set of newspaper accounts of the Ruskin Heights tornado (you may recall that the Rector house was damaged). But she wasn't interested in a romantic relationship.

After graduation I went to the University of Oklahoma to study meteorology at the College of Engineering, a five-year program that required a heavy load of advanced mathematics, physics, engineering, and meteorology classes.

I may have been a little smug about coming from the larger Kansas City to the smaller Norman. After all, Kansas City was home of the National Severe Storms Forecast Center. But I learned quickly that central Oklahoma—when it came to some aspects of weather—was far more advanced than Kansas City.

In Kansas City, we received TV weathercasts from buxom young women, a radio comedian, and a number of other people who had never been in a meteorology classroom. I was immediately impressed by the television weathercasts in Oklahoma City: they had *radar* and they seemed serious about weather.

Just a few weeks after I started my freshman year at OU, on October 5, a tornado struck in Shawnee, in the next county east of where I was living. It killed five people and injured eighty-four others along its path. The after-the-fact news coverage was quite good compared to the norm in Kansas City, but I was troubled by the fact that there was yet again no advance warning. It was a frequent topic of conversation at the OU meteorology department over the next few days. While meteorology had made great progress since the 1950s, I kept being struck by the number of tornadoes that continued to occur without warning, and I wanted to do something about it.

Shortly after the Shawnee tornado, I got a volunteer job at KGOU radio, the new campus station. I did news, weather, sports, and whatever else they asked me to do, as I was eager for the experience.

In February or March, I heard a rumor: WKY-TV in Oklahoma City (the NBC affiliate that had broadcast the first-ever tornado warning) was looking for a part-time meteorologist. Given my radio experience and a couple of brief TV appearances I'd made in high school, I thought I might have a chance, and was excited to get the audition.

I went in for the audition between the six and ten o'clock weathercasts. Frightened wasn't the word for what I was feeling! After a brief tour of the station, I was taken to the studio where I stood in front of a large U.S. weather map. I was given liquid chalk in an Elmer's Glue-All bottle with a wick sticking out of it. They gave me a countdown, "three, two, one!" and pointed at me. The videotape was rolling.

To my delight, a job offer followed, and I was thrilled. I was going to be a *real meteorologist!* It wasn't the glamour or fame of television that attracted me; it was the ability to directly tell people about the weather; there would be no one between me and the audience.

The very next morning, the National Weather Service (NWS) called and offered me an internship in Kansas City for the summer. I had already told WKY that I would take their job, plus I knew I wanted to be in the private sector, as my eventual goal was to be a meteorological entrepreneur; I had no regrets turning NWS down.

When classes ended, I started working at WKY off camera and was trained in the ways of preparing radio and television weathercasts. After about six weeks of rehearsals and training, I was going to debut on the 6:30 a.m. *Farm Show* weathercast. I spent the night before in cold sweats, I was so nervous. Not only did I have stage fright, I kept thinking about what might happen if I made an incorrect forecast or if, God forbid, I missed a tornado. I kept looking at the ceiling, wondering what I had gotten myself into. I don't believe I got a moment of sleep. So when my two alarm clocks went off at 3:15 a.m., I was already wide awake.

I put on my best suit and went into work, plotted the maps, made the forecasts, prepared the graphics, and went out to the small *Farm Show* set. At approximately 6:50 a.m., Russell Pierson, the farm director, introduced me to his TV audience.

The director switched from Camera One to Camera Two. The cameraman pointed crisply at me as the red light on top of the huge blue RCA camera winked on. I began to speak.

The beginning of my career had ended. I was on my way.

CHAPTER EIGHT

STORM CHASERS

THE 1996 MOVIE *TWISTER* IS ABOUT TWO TEAMS of storm chasers. One team is a plucky, ragtag group affiliated with a university that apparently fails to stock hair combs in the university store. The second is an obnoxious corporate group in black minivans. While we never learn why the corporate types are evil, the audience sees the bad guys "get theirs" (an airborne radio tower crashes through the van's windshield) because the guys (and they are all male) in the black vans ignore the advice of the noble university researchers who, in the end, catch their F5 tornado, advance scientific research, and reunite in each other's arms in a happy Hollywood ending.

In the real world of the twenty-first century, there are myriad storm chasers of various motivations: research, journalism, adventure, and, yes, recreation. The only group that is generally well funded is, contrary to the depiction in *Twister*, the university and government chasers. They carry sophisticated equipment, including Doppler radars (Doppler radar senses the winds internal to a storm) mounted on flatbed trucks. The goal of the Doppler on Wheels is to get close enough to the tornado to provide high-resolution mapping of the tornado's wind speed and structure. Earlier storm-chase programs have included launching weather balloons and deploying instruments into a tornado's path (including TOTO, the Totable Tornado Observatory, called "Dorothy" in *Twister*).

Unquestionably, knowledge gained from the thirty-five-plus-year chase program has improved tornado warnings and increased our understanding of severe storms. But there was zero sophistication on a cool evening in Norman in 1972 when the storm-chase program was born.

The quest to improve tornado warnings had been agonizingly slow. In the late 1960s and early 1970s, the National Severe Storms Lab (NSSL) deployed a dense network of weather instruments in central and western Oklahoma, hoping to capture some of the features of the tornadic environment and maybe even a tornado itself. But even after several storm seasons, the tornadoes seemed to have an uncanny knack for eluding the instruments.

In the spring of 1972, we had a routine meeting of the OU student chapter of the American Meteorological Society. The speakers were Dr. Joe Golden of the NSSL and Dr. Bruce Morgan of Notre Dame University. Golden's doctoral dissertation revolved around research on waterspouts (tornadoes over water) in the Florida Keys. Much of the research was conducted by flying airplanes into the vicinity of the waterspouts; Golden and Morgan proposed to do the same for tornadoes. Instead of hoping the tornadoes came to the instruments, the idea was to take the instrumentation to the tornadoes. Morgan wanted to go one step further and drive an instrumented army tank *into* a tornado.

They outlined a Tornado Intercept Program in which volunteer students (no funding, not even gas money) would go out in their automobiles and intercept storms for scientific documentation.

We didn't know it, but the proposed chase program had been received with attitudes ranging from skepticism to full-out opposition from the NSSL, which is why Golden had to ask the students for volunteers; the senior staff at the NSSL didn't find the program scientific enough. Computer modeling was then cutting-edge and an emerging area of NSSL's focus.

When we walked out of the meeting that evening, we were excited by the opportunity to chase storms. There was only one problem: we had no clue how to do it. While Fred Bates, one of the early SELS forecasters, had published some personal observations of tornadoes, there was no known technique for intercepting them because we didn't understand which storms were candidates to produce tornadoes. Nor did we understand (absent a hook echo on radar, and only a minority of tornadoes had hooks) in which section of a thunderstorm the tornado formed. Even if the tornado-bearing thunderstorm had a hook, there was no way to get that information into the field. Remember this was years before cell phones and the remote radar technology we enjoy today. Golden's experience in the Keys wasn't much help, as those waterspouts formed in a different way than Plains tornadoes did, and the wind speeds of the waterspouts were lower, making them less dangerous to chase.

Steve Amburn, my college roommate, and I chased together on a number of occasions, often only to encounter fair weather. This was known as a "blue-sky chase." It was extremely frustrating to drive hundreds of miles and expend a great deal of time for not even observing a single thunderstorm, let alone a tornado. Among the early chasers, the ratio of successful chases (tornado intercepted) to busts was worse than 1 to 20, and a reflection of how little we knew.

My parents thought I was crazy, and they gave me a telephoto lens the following Christmas so I wouldn't have to get "so close" in case I ever found a tornado while chasing. The way things had gone so far, they didn't have much to worry about. I had seen zero tornadoes while chasing.

The original storm chasers took notes, recording visual observations of storms and their locations and times. Most important, they photographed the tornadoes and surrounding storm structures.

* * *

Over the 1972 Christmas break, I encountered Kathleen Rector and she was willing to try dating again; this time it ended successfully, with an accepted marriage proposal a year later. We were in Kansas City making wedding plans during a few days of vacation in late May.

On May 24, 1973, the most important storm chase, possibly ever, occurred when several chase teams intercepted and documented the Union City, Oklahoma, tornado. On this day, for the first time, everything came together: the experimental Doppler radars operated by NSSL in Norman and Cimarron, special instrumentation, and the chase teams in the right places at the right times. This one thunderstorm and the life cycle of the tornado it produced was the subject of literally dozens of scientific papers. For the first time, the internal rotation leading to the tornado was captured and documented. The internal structure of the storm was compared to the damage path caused by the tornado. The visual evolution of the storm was used to help better train storm spotters.

The storm was a watershed event in tornado research and provided valuable information, but it would take years for all of the data to be assembled, evaluated, and published. This is a frustrating aspect of good science: done well, it takes time. The researchers test theories, make discoveries, and then applications-oriented scientists and engineers take the knowledge, test it, and turn it into something useful.

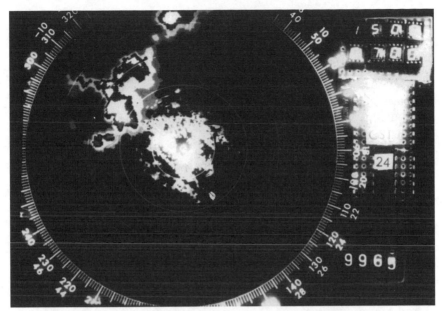

Radar photograph of the Union City tornado. The tornado is contained in the large white echo to the northwest of the radar's location in Norman. Courtesy NSSL.

* * *

On June 4, just days after Kathleen moved to town, conditions looked favorable for tornadoes to develop about sixty miles southwest of Norman, where both Kathleen and I were living. From my perspective, this was an ideal location: we could head southwest and follow the storms northeast back to our home. I thought I would impress Kathleen with my storm-hunting prowess. So we embarked on her first chase. She later said she thought it would be an adventure. Turns out that she was right!

We headed southwest from Norman on Oklahoma Highway 9. We intercepted a towering cumulus cloud in Caddo County, southwest of the town of Chickasha. It rapidly turned into a thunderstorm and looked promising, as we could clearly see rotation starting to develop.

As we drove back northeast toward Chickasha, a wall cloud developed. The chase program had previously identified that this circular lowering and rotating collar of clouds was often a precursor of tornadoes. I was trying to position us ahead and just southeast of the wall cloud so it would safely pass us to the north.

Hail started falling—with almost no rain—and the hailstones got larger with time. As the rotation got more and more pronounced, I was concerned the storm would produce a tornado, so I started looking for a pay phone. I called NSSL and reported what I was seeing. Les Lemon and a grown-up Don Burgess, who had seen his first tornado the night of the Udall storm, were working the Doppler radars.

They were incredulous: the thunderstorm looked very weak and they didn't see any rotation. I told them that something was wrong; clearly the storm was rotating and potentially dangerous.

We left the pay phone and drove a block or so when the danger became much more acute: we were cut off by road construction! Hail continued to batter the car. The wall cloud was getting closer and closer and lower and lower to the ground. It looked like a tornado might form at any second, and we were barely able to move out of its way.

Just before the dangerous cloud passed overhead, we got through the construction and were able to make a dash south to safety. While very impressive to watch, the storm did not produce a tornado at this time. Kathleen was awed by what she was seeing.

I got us back on Highway 9 and started safely chasing the storm from behind. We were keeping up fairly well, but I noticed that storms were rapidly developing to our northwest and headed toward us.

Suddenly, I realized that even though we were headed northeast, we were being rapidly overtaken by the solid line of storms to the northwest. We heard a tornado warning issued over the radio. I felt we needed to break off the chase and try to find shelter—but where? We were in the middle of farms near the tiny town of Blanchard. We saw a farm to our left and I decided to seek shelter there, as the tornado warnings on the radio were getting more and more ominous.

As we drove down the dirt driveway toward the farmhouse, huge hailstones started battering the car accompanied by heavy, heavy rain. We arrived at the farmhouse just as the family was headed for their storm cellar. We introduced ourselves and they invited us to join them. As we were running toward the cellar, I looked northeast toward the original storm (it had stopped hailing) and thought I saw a brief glimpse of a tornado. We went down into the cellar and closed the door.

After about half an hour, we exited the cellar. Night had fallen and the storm had passed our location. We thanked our hosts.

Between flooded roads, darkness, and my lack of familiarity with the area, it was after midnight before I dropped Kathleen off at her apartment. I was afraid she'd have second thoughts about my sanity and was thankful when she didn't break off the engagement. Turns out that with the exception of dodging the flash floods on the way home, which caused a few minutes of anxiety, Kathleen enjoyed the chase. Perhaps this weather stuff wasn't so boring after all!

The storm that we chased in Chickasha did in fact move over Norman. Burgess and Lemon went onto the roof of the Lab just in time to see it drop a tornado on the west side of town. What I had briefly glimpsed as I fled to shelter had indeed been the real thing.

I chronicled this and other chase experiences a year later for *Weatherwise* magazine. Information from the June 4 chase helped document a then-unknown variant of a supercell known as a "low precipitation (LP) supercell." Burgess and Lemon reexamined the radar data from that day and found the reason the storm's appearance on radar did not correspond to what I was describing: LP supercells have a different radar signature than conventional supercells. This knowledge helped lead to a radar technique for accurate warnings. The knowledge of LP supercells was another piece of the tornado-warning puzzle.

It was clear that the chase program was beginning to bear fruit. We were starting, slowly and unevenly, to get ground truth to correlate with what was being seen on the NSSL research Doppler radars; this provided a few of the pieces needed to assemble a rudimentary three-dimensional picture of the tornadic thunderstorm.

Unfortunately, it turned out some rather large pieces were still missing, as I would find out in just a few months.

CHAPTER NINE

TRAGEDY

IN 1973, JUST THREE MONTHS AFTER KATHLEEN AND I were married, I was assigned to do the 6:00 and 10:00 p.m. newscasts for Thanksgiving week when WKY-TV's chief meteorologist was on vacation.

Thunderstorms developed southwest of Oklahoma City on Monday, November 19, in the same general area as the storms Kathleen and I had chased on June 4. The storms became severe, so I went on the air at six o'clock warning of hail and high winds. Because the sun sets early in November, these storms were occurring in darkness.

Since Thanksgiving break began the next day, and because we storm chasers (who were all students) believed chasing in darkness was unacceptably dangerous, there was no one out chasing that evening, so the opportunity for visual reports was much less than it would ordinarily have been in central Oklahoma.

The scattered thunderstorms developed in north-central Oklahoma along with a single large, but not particularly impressive, storm southwest of Norman. The large storm approached the town of Moore, between Norman and south Oklahoma City. I continued to go on the air during prime-time programming to warn of hail and high winds generated by the southern storm near Moore and a couple of storms farther north.

As the southern storm exited Moore, our red emergency phone rang. It was Ray Crooks, the meteorologist in charge of the local National Weather Service; he'd received a law enforcement report that a tornado had moved through Moore, but he couldn't see the storm on his radar. The local NWS radar antenna was located at Will Rogers World Airport, which was so close to Moore that the storm was in the radar's ground clutter. WKY was on the opposite side of Oklahoma City, so Crooks was hoping I could provide some insight.

I rolled my chair a couple of feet to the left in the darkened weather office so I would be positioned right in front of the radar screen. There was the large, green-white radar blob to our southeast exiting Moore. It had no hook, no real indication of a tornado, and I told that to Crooks. But then a thought hit me. "Ray, hang on a second. Let me try something."

Our new WR-100-3 radar had a feature called iso-echo contour, which allowed us to see high-intensity features internal to the storm, a feature our recently replaced radar had not had. I moved the knob to "iso." Inside the blob was something that looked like a black, upside-down candy cane. I could infer from this shape that there might be a circulation inside the storm. I told Crooks what I saw and said we should probably warn people of a tornado just to be safe. He agreed.

I got on the microphone to the control room and said we had to go on the air—*now!* I warned the people of southeast Oklahoma City that a tornado was apparently approaching in the darkness. The National Weather Service's warning came over the teletype a few moments later, causing the warning sirens to sound throughout the city.

Right after I aired the warning, our news anchor George Tomek walked across the hall that separated the newsroom from the weather office and asked, "Mike, did you hear about the barn blown over in Blanchard?" I hadn't. It turned out that a woman had called WKY to report that her barn had been blown over by a tornado. Unfortunately, she had asked for the newsroom rather than the weather office, and the switchboard operator had routed the call accordingly. The newsroom had assumed that I already knew about the tornado.

Because of the holiday, I was working alone, so there was no one to seek out these reports or interface with the newsroom as we usually did. With zero chasers out that evening, the one critical bit of information that could have allowed me to warn the Blanchard-Moore area about the tornado didn't get to me. Worse, it never occurred to me, in the absence of a hook or anything remotely tornadic in the radar depiction of the thunderstorm, to flip the switch to iso-echo until Crooks' call came in.

What I dreaded most had happened to me: A tornado had occurred on my watch and I was not able to warn of it.

I was very uncomfortable with my failure to warn Moore as I continued to broadcast warnings of the tornado moving across southeast Oklahoma City. Little did I know, the situation was about to get much worse.

The storm weakened east of Oklahoma City and the other storms that evening were nothing special, so I turned my attention to last-minute preparations for my 10:00 p.m. weathercast. I told our viewers, at the beginning of the newscast, that the storms had weakened or moved out of our viewing area and reassured them the threat was over. Then George introduced our coverage of the damage—and deaths—in Moore and southeast Oklahoma City.

First up was a report featuring news film taken by WKY's outstanding chief photographer, Darrell Barton. I can describe the images in that brief news report like it was yesterday. Picture Barton, standing in darkness, filming the reporter interviewing officials at a scene of destruction. It was raining lightly and a cold wind was blowing: the front that spawned the thunderstorms had passed, causing temperatures to drop to November-like levels.

A bizarre figure came into view, providing a sharp visual contrast against the darkness: it was a man in a brightly colored basketball uniform. Because of the devastation caused by the tornado, there was great confusion; no one knew where anything, or anyone, was. The man got within range of the officials and asked, loudly, "Where's my family, where's my family?"

The official said, calmly, "Let's find somewhere to talk."

"Where's my family?!"

The official, putting his arm around the man in the basketball uniform, said, "Let's go somewhere where we can talk."

With those words, the viewers knew—as I knew—that the man's family was dead. He had been out of the area playing basketball at the time the tornado hit. Since we didn't have a warning out and the sirens hadn't gone off where he was playing, he had no knowledge of the tornado. When he started home, he was startled to find police roadblocks and debris along the streets of what had once been his community.

The juxtaposition of the frantic man in the basketball uniform outside in the rain, the dark, and the devastation was so incredibly moving it helped Barton win the Television News Photographer of the Year Award for 1974 from the National Press Photographers Association, the highest award a photojournalist can receive. Barton did his job that night incredibly well: seeing the story unfold while standing on the weather set devastated me. I could see the anguish in the man's eyes as he realized the news he was about to receive was going to be the worst he could imagine.

I was sick to my stomach. What I had devoted my very young career to preventing—unwarned tornado fatalities—had already occurred. People had died on my watch.

Of course, there was no way to know whether the man's family had been watching WKY-TV or listening to WKY radio. But they weren't the only ones affected by the storm: five people were killed and forty-six were injured that night. In those days, WKY-TV was so dominant in the Oklahoma City market that it was just assumed that everyone was watching us during severe weather, which is part of why I took the failure so personally.

With difficulty, I got through the regular weathercast a few minutes later, then drove home, deeply depressed and upset. Kathleen consoled me as I poured my heart out to her. Feelings of sorrow for the victims, great disappointment, immense failure—it was all there.

To this day, the memory of that night more than thirty years ago is uncomfortable and painful. It is difficult to write about. I hope, somehow, the man was able to rebuild his life.

The thunderstorm that caused the tornado that struck Moore was in several ways similar to the storm that produced the Union City tornado: neither presented themselves as obvious tornado-producing storms. But six months just wasn't enough time for the knowledge being generated by studying Union City to be of benefit. Besides, even if we had known that the storm types were similar, it might not have mattered. We needed Doppler radar to correctly diagnose that particular storm and there were no operational Dopplers in those days.

It would be many years before Doppler radar made it into widespread use.

* * *

One thing about operational meteorology is that the weather never stops; there will always be another forecast, another opportunity to get it right.

Three days later on Thanksgiving morning, it looked like a rare November snowstorm was going to hit Oklahoma City. I forecast a chance of light snow on the six o'clock evening weathercast. Later in the evening, additional information flowed into the weather department. A strong low-pressure center located near Albuquerque was moving nearly straight east with more atmospheric humidity than originally predicted. So at ten, I went on the air and forecast that snow was almost certainly to occur during the night and that it would accumulate, probably a couple of inches. Oklahoma City is far enough south that two inches of accumulating snow is a very big deal.

I was curious what the "competition" was going to forecast. The ABC affiliate, KOCO-TV, had a delayed start to its 10:00 p.m. news because of a late-ending football game. KOCO's meteorologist, Fred Norman, began his weathercast by saying, "If you were watching the Brand X station a little while ago you saw them forecast two inches of snow tonight. I am here to tell you that the low-pressure system [a different system than I was watching] has passed us by and there will be no snow for us tonight or tomorrow."

No sooner had Norman spoken those words than our red phone was ringing. It was our news director, Ernie Schultz, very upset with me for forecasting snow. Of course, Schultz was not a meteorologist, but he'd watched KOCO and KWTV, and no one else was forecasting snow. I was the odd man out *and* only twenty-one years old, while the competing meteorologists were considerably older and more experienced. Schultz was acutely aware that I had failed to forecast the tornado just three nights before. He let me know that I had better be right about the snow.

I went home tense, to put it mildly. Kathleen was in Kansas City celebrating Thanksgiving with our families, so I couldn't talk it out with her. I settled down to watch Tom Snyder's *Tomorrow Show*, but I kept peeking out the window during each commercial break to glance at the sky. I tried going to bed but I kept getting up, pulling back the curtains to see whether snow had begun falling. Nothing. Exhausted after a terribly difficult week, I finally fell asleep sometime after 3:00 a.m.

I awoke on Friday morning, and seconds after reaching conscious-
ness, with no real thoughts as yet going through my head, I noticed
the room looked odd. The ceiling was unusually bright. While the
curtains were closed, there was about a four-inch gap between the
curtain rod and the window, allowing unusually bright light to reflect
off the ceiling. Could it be? I jumped up and swept open the curtains:
Under a deep blue sky, bright sunshine was reflecting off the daz-
zling snow cover. I could see the snow was exactly two inches deep.
I was so excited that I called Kathleen in Kansas City (long distance
phone calls were expensive in those days), practically jumping up and
down as I spoke on the phone. The successful snow forecast didn't
even come close to making up for my failure to forecast the tornado
earlier in the week, but it helped. I regained some confidence. However, when I ran into Schultz later that day, he didn't congratulate me
for the correct forecast nor did he apologize for the previous night's
irate phone call. I never held it against him as the imbalance in the
attention paid to meteorological successes versus failures seems to go
with the territory. The fate of the forecaster is well summed up in this
anonymous poem:

OLD FORECASTER'S LAMENT
And now among the fading embers
These in the main are my regrets
When I am right no one remembers
When I am wrong no one forgets.

CHAPTER TEN

FUJITA

WE OFTEN ASSOCIATE DETECTIVES WITH POLICE
work. A police officer at a crime scene gathers fingerprints, collects
hair samples and tests for DNA, and takes photographs to determine
who might have perpetrated a crime.

There are detectives in other fields, too. The medical pathologist
peers through a microscope in a lab to determine what might be causing a patient's unusual set of symptoms.

In meteorology, many experts were doing detective work in the
1970s, 1980s, and 1990s. But there was one weather detective extraor-
dinaire: Ted Fujita. Ted's unique approach yielded breakthroughs that
have benefited just about everyone who has ever traveled by plane or
experienced a major windstorm.

Born in Kitakyushu City, Japan, in 1920, young Fujita developed a
fascination with maps as tools for relating data he gathered through
self-taught surveying techniques. In college, he studied mechanical
engineering and physics. He was ahead of his time in developing

manual mapping techniques that today would be considered geographic information display (GIS) technology. Young Fujita used his maps and tools to survey the aftermath of Hiroshima and Nagasaki to determine the height at which the bombs detonated. He created three-dimensional maps of volcanic craters and topographic maps of the areas around his home.

On July 17, 1948, after hearing thunder in the distance, Fujita took a pencil and notebook and went outdoors to observe the storm. He mapped thirty-three lightning strikes and the storm's motion and realized he'd been bitten by the weather bug. Later that same year, he surveyed his first tornado, and in 1949, he surveyed his first hurricane, both in Japan.

In 1945, the U.S. Congress passed a law requiring the Weather Bureau to study thunderstorms. The impetus for the law was commercial aviation accidents in 1940 and 1943. Hearings were held and appropriations were made; the Thunderstorm Project was born.

The Thunderstorm Project, directed by the University of Chicago's Horace Byers, came at an advantageous time: there had been a number of recent advancements in meteorology (the discovery of the jet stream, for example) and aviation (improved and more reliable aircraft) resulting from World War II. These advances made it possible to focus investigation and gather better data than would have been possible just five years earlier.

In 1946 and 1947, in Florida and Ohio, the project launched aircraft and balloons to gather data, installed fifty-five weather stations, and used radar to probe the airflow in and around cumulus clouds and thunderstorms. The Army Air Force provided aircraft to penetrate the storms at various altitudes. The data was collected and analyzed. Research reports were published in 1948 and 1949.

In 1950 Fujita came across a copy of one of Byers' research papers and was fascinated. He wrote to Byers and enclosed a copy of a scientific paper of his that analyzed a Japanese thunderstorm.

Byers answered Fujita's letter on January 30, 1951, and Fujita called it "the most important letter I received in my life." Byers complimented Fujita on his work, relating that Fujita had unknowingly verified some of the findings of the Thunderstorm Project, and included a copy of *The Thunderstorm*, the project's final report.

The letter from Byers to Fujita contained a prescient sentence, "In particular, you deserve credit for noting the importance of the thunderstorm downdraft and outflowing cold air."

Fujita received his doctorate in 1953. That same year, Byers invited him to Chicago for two years of research. Fujita accepted; he would never again reside in Japan.

One of the critical challenges preventing improved storm warnings was that most important storms were much smaller than the spaces between weather stations. For example, the eye of a hurricane can be as little as ten miles in diameter. But in most areas, weather stations are fifty miles or more apart, so getting the dense measurements needed to improve our understanding was a major challenge. Fujita was a coauthor of a 1956 paper, *Mesoanalysis*, which made public his techniques for filling the space between the conventional meteorological data.

Without Fujita's techniques, increasing our knowledge of thunderstorms and similar small-size meteorological events during the 1950s, 1960s, 1970s, and 1980s would have been nearly impossible.

On June 20, 1957, a violent tornado struck the Fargo, North Dakota, area, killing ten people and injuring more than one hundred. It was on this storm that Fujita performed his first study of a U.S. tornado, which raised the bar for every future study.

Consider this: There was no radar. There were no weather satellite images. The only weather station in the area was at the Fargo Airport, miles away, and the second nearest station was tens of miles away. Studying the genesis of a tornado, the size of which was measured in hundreds of yards, seemed an impossible task.

But Fujita wasn't an ordinary meteorologist. He had a creative perspective and a mind that viewed the world in four dimensions: the north/south dimension, the east/west dimension, the vertical dimension (altitude), and time. In order to obtain the data he needed, Fujita painstakingly worked with Fargo's WDAY-TV to gather 150 photographs of the tornado taken at various times and various locations in an age where tornado photographs were a rarity.

Having worked with Fujita on a study of the Hesston, Kansas, tornado in 1990, where we had the advantages of videotape and aerial photography, I can hardly imagine the amount of meticulous work needed to assemble the 150 still pictures into a coherent whole from which conclusions could be drawn. It wasn't enough to view the photos themselves; it was necessary to learn the location from which each photo was taken and the time it was taken so the chronological sequence of cloud structures and features could be assembled. In 1957, no other meteorologist would have even known how to *begin* this type of study.

To perform his analysis, Fujita almost single-handedly created the art of meteorological photogrammetry. He triangulated the photographs from different locations in order to track the evolution of the storm's features. From his work, he was able to synthesize an until-then unknown view of the development of a tornadic thunderstorm. Fujita was the first to discover the "wall cloud" that we now know to be a frequent visual precursor of a tornado. Unfortunately, the wall cloud was more or less forgotten by meteorologists until the chase program of the 1970s reconfirmed Fujita's work. That knowledge is used by storm spotters today to help meteorologists provide warnings about tornadoes more quickly.

Fujita worked with Byers and the U.S. Weather Bureau to publish his findings in a 45-cent government printing office pamphlet. It sold out. Most of the 3,000 copies published were purchased by Fargo-area residents.

Fujita's interests and abilities blossomed. In 1965, a swarm of tornadoes struck the Midwest on Palm Sunday. Fujita mapped and studied what meteorologists call a tornado outbreak.

One of the mysteries of tornadoes was why two buildings might be destroyed while a similar building between them remains relatively untouched (I had first noticed this peculiarity at age five while traveling through devastated Ruskin). The conventional explanation from meteorologists was that the tornado "was going up and down." By studying the marks tornadoes left in plowed fields, Fujita discovered the "suction spot," a mini-vortex that rotates around the primary tornado funnel. The suction spot (now generally called "suction vortex") causes the most intense damage.

Another mystery unlocked.

Ted Fujita. Photo courtesy the University of Chicago.

In the 1960s and 1970s, the federal government was increasingly interested in tornado wind speeds, in part for nuclear power-plant design. Dr. Robert Abbey of the Nuclear Regulatory Agency turned to Fujita, who developed the Fujita Scale that allowed meteorologists, for the first time, to classify tornado damage and correlate it to wind speeds. This research has been used to better understand the structure of tornadoes and to design buildings and shelters better able to withstand them.

The Fujita Scale is one of the few pieces of meteorological jargon that has made it into the vernacular, courtesy of the movie *Twister*. Throughout the movie, moviegoers hear about F-4s and F-5s, which are the highest-intensity tornadoes on the scale. Less than one half of one percent of all tornadoes are F-5s, but they are the ones that are most likely to cause multiple deaths as well as the most intense damage.

As Fujita's reputation grew, so did the envy of scientists who lacked his analytical gifts. Fujita's analysis techniques were called unconventional. A small chorus of doubters arose. After all, no one had ever *seen* a suction vortex. No wind measuring device had ever survived a direct hit by a strong tornado, so who knew whether the Fujita Scale wind speed estimation technique was any good? This envy peaked with Fujita's discovery of the "downburst," which I cover in Chapter 13.

Nine years after the 1965 Palm Sunday tornado outbreak in the Midwest, an even worse tornado outbreak occurred on April 3–4, 1974, killing 319 people. Included in the outbreak were tornadoes that struck Louisville, Kentucky; Cincinnati, Ohio; Xenia, Ohio; Huntsville, Alabama; and other locations.

Fujita led a team of meteorologists that included my atmospheric physics instructor that semester, Dr. John McCarthy, on a research expedition. The goal was to survey the path of every tornado. Fujita's team pooled their information with the National Weather Service study of the same event. The team produced a map of each of the 148 tornado paths that, put end-to-end, spanned 2,200 miles! They rated each tornado path segment with a Fujita Scale number.

Fujita returned to the University of Chicago after the fieldwork. McCarthy came back to teach our class. When the surveying was over, it took the team ten months to analyze their results.

While the warnings issued during the 1974 tornado outbreak saved many lives, a post-event study by the National Weather Service revealed a number of serious flaws in the warning system: The 55-word-per-minute teletypes were too slow to carry such a deluge of warnings, especially since there was no system to edit out non-essential messages. The communications system was still based on the punched paper tapes that were being relayed by hand from circuit to circuit. The system was so overwhelmed that some warnings were being transmitted even after they had expired, taking up valuable communications bandwidth that crowded out critical warnings still in effect. As a result of the post-storm investigation, an expansion of the National Weather Service prototype radio network was recommended.

Radar problems during the outbreak were also studied. And while the early editions of geostationary weather satellite images were judged to have been helpful to the forecasting process on those two April days, the images were inadequate to assist in the warning process—higher resolution and more timely data was needed, both at the NSSFC and in the local forecast offices that issued the warnings.

The National Weather Service final report on the storm was remarkably candid and led to a number of new warning procedures as well as early plans to upgrade the radar and warning networks. Further, the study was a key element in follow-on research that improved the dissemination of tornado warnings as well as added to our understanding of how major tornado events develop.

While the National Weather Service worked on improving the national warning system, Fujita's team focused on the scientific characteristics of the storms themselves. Suction vortices, tornado "families" (i.e., sequential tornadoes from a single supercell), and other useful scientific results flowed from their research.

CHAPTER ELEVEN

THE DAY TV WEATHER
GREW UP

BACK IN OKLAHOMA AFTER SERVING ON FUJITA'S
survey team, Dr. McCarthy returned to teaching our class. Speaking
emotionally for about twenty minutes, he told us about his experiences
surveying the tracks of the tornado "families," and the devastation he'd
seen. Just before he turned his attention to the day's lesson, he added,
"Mr. Smith, please stay after class. I want to talk with you."

I had no idea what Dr. McCarthy might want to talk about, so I
assumed that I was in trouble. It was my final semester before graduation, and I'd been struggling with atmospheric physics. This course
was not typical meteorology; it dealt with topics like atmospheric
light refraction. I'd always had a good instinct for weather, but these
topics didn't come naturally to me.

After the day's lecture ended and the other students left the classroom, McCarthy sat me down with a concerned and troubled look on
his face.

"The warning system just broke down," he said.

Based on the scuttlebutt in the meteorological community while McCarthy had been gone, I was not surprised to hear him confirm what many had suspected.

"The warning system just broke down," he repeated. "It couldn't keep up. The Weather Service offices couldn't get the warnings out fast enough. The weather wire fell far behind. The TV stations' warnings were often ineffective." There was a pause, then he added, "You need to be thinking about what you would do if a major tornado outbreak occurred here in Oklahoma."

What *would* I do if a major tornado outbreak occurred in Oklahoma? Even though tornado outbreaks are infrequent (every few years or so), I ran through scenarios of what I'd do if a major tornado outbreak were to occur in our viewing and listening areas. In all, I gave it several hours' thought because I knew McCarthy was right: the normal warning process would break down in a major outbreak. I had a plan in mind, but would it work? I was about to find out.

Eight weeks later, an unusually strong storm began moving toward Oklahoma from the northwest. I was now the de facto number-two man in the WKY-TV rotation. When meteorologist Jim Williams took vacation time, I was generally the one who filled in. Williams was off Friday, June 7, and I was assigned to do the 6:00 and 10:00 p.m. weathercasts in addition to my normal weekend broadcasts.

All of my scientific training and every instinct I had told me this storm was going to be a big deal. So at 10:00 p.m., and for the first time in my career, I told viewers that a major severe weather outbreak, including the possibility of tornadoes, was possible the next day and that they should pay close attention to the weather.

Prior to leaving for the night, I went upstairs to the WKY radio station (which, unlike TV, was on the air twenty-four hours a day) and left word with the overnight newscaster and disc jockey to call me if anything came up during the night.

Once home, I didn't sleep very well. Even though the sky was clear when I had left the office, I knew the phone would ring well before dawn.

At 3:30 in the morning on June 8, the phone did indeed ring. Severe thunderstorms were developing southwest of Oklahoma City. I got up, dressed, and headed in. For the rest of the morning, I broadcast severe thunderstorm and flash flood warnings on WKY radio and, as soon as it signed on for the day, TV. There were literally dozens of warnings, and between the two stations, I was on the air almost continuously.

By 11:00 a.m., things had calmed down and the storms had moved out of our viewing area, so I went home for some lunch and a short break. I knew that Round Two was going to be far worse.

I was back in the weather office shortly after noon. A National Weather Service tornado watch was issued moments after I arrived, and by one o'clock, the first radar echoes were starting to form southwest of Oklahoma City. I went on the air announcing the watch and advised people to stay tuned   it was going to be a dangerous afternoon.

Severe weather technology in 1974 was still primitive. There had been little improvement in weather radar displays. Radars still produced glowing white blobs on cathode ray tubes (CRTs) in darkened rooms. There was no direct geographic information, simply blobs representing thunderstorms over concentric range marks surrounded by an azimuth (direction) ring. In those days, meteorologists could be heard shouting, "Hook at 223 degrees, 42 miles!" Of course, this was meaningless to the public. For TV and radio, we had to convert the information into something people could understand.

I did have one important new tool: WKY-TV had installed the first-ever computerized weather radar for television just a couple of weeks before. Unlike the system we had had during the Moore-Blanchard tornado, this system had a primitive computer that displayed gray-toned echoes that gave a view of the internal structure of the storm with each sweep of the radar. The advantage of this computerized radar was that it allowed the most intense storms to manifest themselves more clearly and quickly, allowing me to focus on the ones that were truly dangerous. I didn't have to switch to iso-echo like I had done—after the fact—just seven months earlier. This would be the first real test of the new system.

The storms were heading right for Oklahoma City. From what we'd learned from the storm chase program, the southernmost storm in a line of this nature was most likely to produce a tornado.

And it was the southern storm that was heading our way.

I went on the air to warn viewers and listeners of the tornado danger, and I described the developing hook echo to add credibility and immediacy to my predictions.

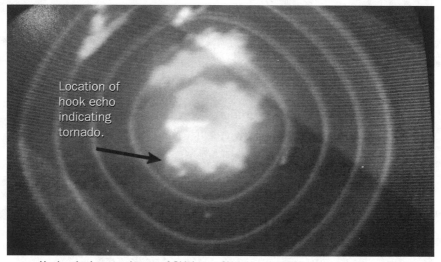

Hook echo just southwest of Oklahoma City, June 8, 1974.

The TV station was on the far north side of the city. Our radar had a good view, as the storm was outside the radar's ground clutter. Realizing the tornado was in the ground clutter of the NWS radar at Will Rogers Airport, however, I called the Oklahoma City NWS office to let them know of the danger. My former college roommate, Steve Amburn, answered the phone.

Amburn, who had just started his career as an NWS intern meteorologist, informed the senior meteorologists on duty, Jerry Cathey and Don DeVore, that I had reported a developing tornado. Because a large aircraft hangar was immediately west of the NWS office, their views were blocked. Cathey and DeVore decided to walk outside, around the hangar, to look for the developing tornado, as the storm was so close their radar was useless.

Cathey and DeVore knew they were in serious trouble when the roof suddenly came off the hangar. Running back to the office, pelted by rocks, they had trouble opening the door in the high winds. Just as the building's windows blew out, they got the door open, shouting, "Issue a tornado warning! Issue a tornado warning!" Their office suffered a gas leak and had to be abandoned. Ray Crooks, the Oklahoma City NWS chief meteorologist, called WKY and told us to "take over."

Jim Williams, who had come in just a few minutes earlier to assist me, took the call from the NWS. I was really glad for his help because things were changing by the minute as the storms continued to grow in number and intensity. The tornado that hit Will Rogers Airport continued into the south part of Oklahoma City. That storm, alone, would have been enough to handle, but now a half dozen other severe thunderstorms, many of which were candidates to also produce tornadoes, were requiring attention.

I did the calculations in my head and realized that if we immediately put a camera out the back door and aimed it to the southeast, we might be able to capture the tornado live on the air. I asked the camera crew whether they could move a camera out to the back loading dock, which was on the south side of the building. They managed to roll a heavy studio camera outdoors onto the slanted loading ramp. When the camera was ready, we took the shot live. There it was! The tornado's rotating wall cloud was clearly visible. I narrated by explaining that a wall cloud was a precursor to a tornado. The rotation in the wall cloud continued to tighten and a funnel began to form as though right on cue. I reminded viewers of how serious the storm was and urged them to take shelter. Moments later, the funnel cloud traveled around the corner of the building, out of the camera's range. We later learned the funnel cloud reached the ground and became a tornado. It caused damage while we were live, but the camera wasn't able to pick up any level of clarity through the rain and haze.

It turned out this was the first time—ever—that a tornado had been broadcast live on television. In addition, we were using the radar to pinpoint the location of the tornadoes and to *tell people where they were headed.* The short-term forecast of a tornado's path was unheard of in 1974. For example, while we were televising the live tornado, I made comments along the lines of "eastern Oklahoma City, Luther, Harrah, Spencer, and Wellston are in the path of this tornado and residents of these areas should take shelter immediately!"

The other supercells were now starting to produce tornadoes in central Oklahoma. Larry Brown, our other meteorologist, had joined Williams and me in the office. All hands were needed to deal with the flood of data and calls we were receiving. At some points during the afternoon, four tornadoes were on the ground simultaneously in our viewing area. In addition to the tornadoes, there were too many severe thunderstorms with large hail and damaging winds to count.

The warning system should have broken down, as the density of tornadoes in our viewing area had reached a level similar to that of the April 3 tornado outbreak in the states to our east. And remember, the Oklahoma City NWS was temporarily out of service. The Tulsa NWS office had an older radar and could not adequately view all of the central Oklahoma storms. But despite these challenges, things went incredibly well.

Why didn't the system break down? Because John McCarthy urged me to plan for what I would do in a major tornado outbreak.

Rather than following the "all warnings are equal" paradigm that was in place in 1974, Williams, Brown, and I focused on the truly dangerous storms. This shift in thought and procedure allowed us to save lives.

The three of us worked as a team. Williams handled most of the radio reports, and Brown organized the wire copy and the police and spotter reports; they passed all critical information along to me. I alternated my time, every five minutes or so, between the darkened radar room gathering data and the brightly lit TV studio giving on-air advisories. While I was scanning the radar, I also monitored the information from Williams and Brown. The three of us carried out our roles for hours.

Oklahoma City was hit by tornadoes four additional times, for a total of five times in one day, a record for the most times any city has ever been hit by tornadoes in a twenty-four-hour period. That record still stands. In addition to the damage Oklahoma City sustained, a number of towns throughout central Oklahoma were in ruins.

By evening, the tornadoes had moved into northeast Oklahoma. Normally, we would have ceased our coverage, but the Tulsa area had been hit hard by tornadoes and flash floods. We got a call from a cable TV operator saying all of the Tulsa TV stations were off the air and that he was going to temporarily televise WKY-TV, if we would cover the northeast Oklahoma tornadoes. We did. This extended our coverage for several hours.

Finally, around midnight, it was over. I was hungry and exhausted, and my eyes were severely strained from the abrupt lighting changes between viewing the radar in the dark weather office and broadcasting under bright floodlights. We were getting conflicting reports of damage and fatalities. I left the station with a sense of pride about my efforts mixed with concern about whether I had done enough.

When it was all totaled, twenty-two tornadoes had occurred in Oklahoma that day, causing sixteen deaths. We had somehow managed to get a warning out in advance of *every* tornado that occurred in our viewing area.

And there were *zero* deaths sustained in our viewing area.

All of the fatalities were in the Tulsa television market, which didn't have live radar and some of the other tools we used in Oklahoma City. But I also believe that Dr. McCarthy's advice had paid off. By planning ahead and preparing how to handle a tornado outbreak, our warning system didn't break down.

I was completely unprepared for what happened next.

Over the next few days, more than seventy-five letters and postcards came into the station, complimenting me on our coverage. Almost all were highly emotional, along the lines of, "Thank you for saving my family's life. You told us to take cover and we did, and a few minutes later the tornado blew our home away." It was completely overwhelming reading letter after letter.

A letter to the editor thanking the WKY-TV weather team was published in the *Daily Oklahoman*. The *Oklahoman* even ran an editorial cartoon complimenting our effective television warnings. Considering the *Oklahoman* was a frequent critic of television in general, this was high praise, indeed.

Courtesy of the *Daily Oklahoman*.

Word of what we had accomplished on June 8 quickly spread via television news consultants, hired by television stations to improve their news programs in order to optimize their ratings and revenue. It seemed, practically overnight, that bona fide meteorologists and top-of-the-line radars were popping up in TV market after TV market. Across the United States, the reign of puppets, comedians, and big-busted "weathergirls" (as opposed to meteorologists of the feminine gender) as forecasters was largely over within the next three to five years.

* * *

Kathleen and I moved to Wichita in 1975 when I became chief meteorologist for KARD-TV, the NBC affiliate, where I created a professional weather department and upgraded their radar. While the new radar gave us state-of-the-art storm detection capabilities, our on-air visuals were still black-and-white displays that had been frustrating TV meteorologists since the 1950s.

In 1976, I was approached by Technology Service Corporation (TSC) of Los Angeles, who said they had the capability to create color radar. Would I be interested in working with them on the project? Would I!

A few months later, a metal box about the size of a suitcase arrived in Wichita. Inside was hand wiring so intricate and well constructed that it was a thing of beauty. To this day I have never seen a system as well designed and well built as TSC's radar converter, used to convert the radar analog signal to a digital signal. It was still working flawlessly when it was retired nearly twenty years later.

I knew that the signal processing could be done to turn radar into color. TSC had developed this technology. But what exactly did that mean? How would we create color radar images that viewers would understand? It was my job to turn the technology into something useful.

The first step was to create maps (far nicer than the primitive black-and-white TV radar maps) that would make it easy to relate the location of the storms to where viewers lived. The mathematics and hand plotting involved to convert a Mercator map of the Wichita viewing area into a form that could be digitized for the computer inside the TSC device took about a week.

For the second step I answered the question "What colors should the weather be?" Here's what I decided. Snow produces only weak radar echoes. Snow is cold. Blue is a cold color. We made light echoes light blue, which would be suggestive of snow in winter and very light rain in summer.

People associate light to moderate rain with plants "greening up." So we made that intensity of rain green. Very heavy rain constitutes a cautionary situation, in that it is difficult to drive in, it cancels outdoor events, and if it persists long enough can cause flooding. We chose cautionary yellow to designate heavy rain.

The worst storms, with torrential rains and/or large hail were the ones to look out for. Red is a "danger" or hazard color, so those storms we made red.

These colors may seem obvious today, but they certainly were not in July 1976. We were creating a never-before-seen system. It is fitting that the most revolutionary technology in the history of TV weather coincided with the bicentennial of the American Revolution. And just as colonists fervently clamored for independence, so our viewers took to the color radar.

Our cross-town competitor KAKE-TV couldn't purchase the new color radar fast enough. And within a year, dozens of TV stations across the United States had colorized their radar, and by 1980, black-and-white radar had disappeared from the scene.

CHAPTER TWELVE

ST. LOUIS AND THE HOLIDAY WEATHER HOTLINE

KARD-TV'S WEATHER PROGRAMMING WAS A HUGE success. But by the spring of 1979, Kathleen was feeling a little restless in Wichita, and I was looking for a new challenge. I wanted to see whether my forecasting style would work in a large market.

I received a phone call from Rabun Matthews, the news director at the ABC affiliate in St. Louis, KTVI. Matthews had been brought in from Seattle to completely overhaul KTVI news, and he wanted to know if I would be interested in interviewing for the chief meteorologist position and setting up a weather department there.

KTVI was known throughout the industry as one of the biggest basket cases in broadcasting. In the late 1970s, ABC was the number-one network nationally, but as a result of years of inconsistent strategy, KTVI was often number four in what should have been a three-station race (KPLR, an independent station, presented a late newscast in addition to the three major network affiliates).

I met with Matthews and the general manager, Ralph Hansen. They convinced me that they were going to do everything they could to bring up the ratings; ABC had threatened to yank their affiliation if they didn't.

Here was my chance to see whether my ideas would work in a big market: at the time, St. Louis was the number twelve market in the nation. St. Louis was also in the middle of the country, so it had blizzards, tornadoes, flash floods, and thunderstorms. We decided I would take the job, and after giving KARD (which had been very good to me) six weeks' notice, I started at KTVI on September 1, 1979.

The critics' reviews of my debut forecast in St. Louis were mixed. The *Globe-Democrat* liked what they saw, but the *Post-Dispatch*'s Eric Mink was critical of my "over-electronification" of the weather. But overall I thought I'd started off fairly well.

I knew we had to think outside the box to draw some much-needed positive attention to KTVI. I proposed we implement a new program—the Holiday Weather Hotline. With two meteorology schools in the St. Louis area and access to the first easy-to-use computer databases in meteorology, I thought we could hire students to staff phone banks in the days before Thanksgiving and Christmas. A company in Oklahoma City, Weatherscan, had developed an electronic database that allowed us to rapidly call up data for anywhere in the United States. Holiday travelers could call our phone banks to learn about the weather in the area they were traveling to.

KTVI allowed me to try the program, and they promoted it with full-page newspaper ads. We fielded more than 20,000 calls from holiday travelers, far exceeding our expectations. We were slowly getting the market's attention and we saw a very slight uptick in the November ratings.

Before my joining KTVI, the St. Louis TV stations more or less ignored storm warnings that occurred at inconvenient times. Our key to success was to change that outdated storm-coverage paradigm— after all, more people had been killed by tornadoes in St. Louis than in any other major city, a record that still stands today. My perception was that St. Louis viewers were hungry for quality weather coverage, but no one had yet offered it to them. I vowed we were going to be on the air first, and with the best coverage, regardless of when a weather hazard happened to present itself.

The KTVI management viewed the 1980 Winter Olympics as a major showcase for our still-new revamped news product. We were hoping people who were watching ABC for Olympic coverage would sample our news.

One night during the popular women's figure skating competition, some fierce thunderstorms developed over the Illinois portion of our viewing area. Damaging hail and winds were likely to come from these storms. I had already alerted the crew that cut-ins were a possibility. However, because this was not a life-threatening tornado or flash flood situation, I told them to wait until the skater had finished her routine before interrupting the programming. That way we could relay our information and be back to Olympic coverage by the time the judges were ready to show her scores.

It went off like clockwork. We demonstrated we were serious about covering dangerous or damaging weather whenever it occurred, but we didn't preempt anything of substance. Our commitment to excellence was intact.

That year, St. Louis experienced several dramatic weather occurrences. One of them, a type of widespread, intense thunderstorm-related windstorm now known to meteorologists as a "derecho," produces winds similar to a Category One hurricane. After reviewing incoming weather data, we recognized that a major windstorm was going to sweep across our viewing area. Our warnings for this storm were far ahead of our competitors'. As the winds were sweeping through St. Louis, I told the studio crew to aim a TV camera at a set of electronic LED readouts from weather instruments we had on the roof. As I continued to broadcast live warnings, we showed the wind speed, which suddenly took a jump to more than 60 miles per hour. We even caught a gust of 81 miles per hour live. This may not sound like much now, but in 1980 these things had never been done. As power failed across the city and high winds downed countless trees, we succeeded in providing the city with excellent advance warnings.

We received three thank-you letters after that storm, one from a woman living outside St. Louis who had had no inkling there was a storm in the area until she saw our warning. As she ran outside to get her children, the storm became visible advancing over a row of trees to the west of her home. She and her children got to the basement just as the storm struck. When they later emerged from the basement and viewed the front yard, the woman saw a large tree lying across the exact spot where her children had been playing.

KTVI was moving up in ratings and profitability, and morale was high. But in late winter 1980, our station was purchased by Times Mirror. Unfortunately, this change in management was not a positive one. Even though the February 1981 ratings showed us to have achieved number-two status, the highest news ratings KTVI had ever had, internal politics made any real progress next to impossible.

At about that time, Don Sbarra, the president of KARD-TV (now known as KSNW-TV), offered me the opportunity to return to Wichita but with an interesting twist: I would not be working for the station. The station would assist me in starting my own weather company. This was my dream.

Kathleen and I would have liked to remain in St. Louis, but after much thought and discussion, we decided to return to Wichita to fulfill my dream of starting my own company.

As we prepared to move back to Wichita, KTVI received the July ratings. We had done it! KTVI, for the first time in its history, was number one. Not only that, but the *Post-Dispatch* took a poll of television personalities in St. Louis, and I was named the most popular weather personality, a fact confirmed by KTVI's own audience research. It was a nice note on which to leave.

CHAPTER THIRTEEN

THE MICROBURST MYSTERY

We're gonna get our airplane washed.

What?

We're gonna get our airplane washed!

Approach, Delta One-Ninety-One is with ya at 5 [thousand feet].

To the untrained ear, it sounded like three men making small talk. But those words, the words to follow, and their subtext would comprise one of the most closely analyzed conversations in history.

Lightning coming out of that one.

What?

Lightning coming out of that one.

Where?

Right ahead of us.

Four microphones, one at each of the pilot seats and one to record any ambient sounds, were in the cockpit of the Lockheed L-1011 jumbo jet. But it was doubtful the three pilots ever thought about the whirring mechanism recording their voices on a metal ribbon in the very back of the aircraft.

The location of the orange-colored "black box," officially known as the Cockpit Voice Recorder or CVR, is chosen to ensure the maximum chance of its survival in case a plane crashes. A second black box, the Digital Flight Data Recorder (DFDR), records the plane's altitude, angle, whether the landing gear is up or down, and numerous other details that would be critical in an investigation. By law, all commercial airliners carry these devices.

First Officer Price: We're gonna get our airplane washed.

Captain Connors: What?

First Officer Price: We're gonna get our airplane washed.

On August 2, 1985, Delta Airlines' First Officer Rudolph Price was flying the final approach to Dallas-Fort Worth (DFW) International Airport.

Price: Stuff is moving in . . .

Connors: Tower, Delta One-Ninety-One heavy, out here in the rain. Feels good.

It sounded as if Price, Captain Edward Connors, and Second Officer Nick Nassick were casually discussing a small rain shower developing between their aircraft and the runway.

A second or two later, the small talk turned to urgent concern.

Delta 191 was in the final moments of a flight from Fort Lauderdale to Dallas. After a stopover in Dallas, the plane was scheduled to go on to Las Vegas. Price was flying this particular leg of the flight. Commercial airline crews usually fly together for a few days at a time, with the captain flying the first leg of the day, then alternating with the first officer until the day's flying is complete. On an L-1011, the second officer is a fully qualified air transport pilot but does not routinely fly the plane.

Behind the three pilots were 152 passengers and eight flight attendants. The passengers were strapped in for landing, and the flight attendants had just taken their seats.

There was some light turbulence on this particular approach. When the temperature reaches 101°F, as was being reported at the DFW weather station, the ground heats the lowest layer of air. Because this near-ground air is hotter than the overlaying air, it rises like an invisible hot air balloon. As air ascends, it cools. And cool air can hold less moisture than warm air.

So as the air ascends, the humidity rises, until it reaches nearly 100 percent and a cloud forms. On this early August evening, the sky featured the cotton-ball-like cumulus clouds typical of a Texas summer sky.

Delta 191 pierced a few of those cumulus clouds as it descended to its final approach. The plane received a few solid turbulence jolts while it was briefly in the clouds, but the turbulence was nothing out of the ordinary.

Now under the cloud base, the crew could see what appeared to be a light rain shower between them and the runway. Other planes were flying through the shower and landing normally. But once Delta 191 entered the rain shower, all hell broke loose.

Connors: Watch your speed.

[Sound of rain striking the fuselage begins.]

Connors: You're gonna lose it all of the sudden, there it is!

Connors: Push it [the throttles to increase engine power] up, push it way up.

Connors: Way up!

Nassick: Way up!

Connors: Way up!

Connors: That's it. [a note of relief in Captain Connors' voice]

Connors: Hang on to the sonofabitch!

Whoop whoop pull up, whoop whoop pull up [Flat, unemotional computer-generated voice triggered by the aircraft's ground proximity alarm]

Connors: Toga! [The "take off/go around" command to put the aircraft into "go around" mode and abort the landing.]

Whoop whoop pull up

Nassick: Push it way up!

For eight years there had been a fierce controversy within the meteorological community. Ted Fujita, along with his mentor, Horace Byers, had published a paper in 1977 describing an unknown meteorological phenomenon, something Fujita called a "downburst"—air that descended violently from a thunderstorm, causing high winds at the earth's surface. This phenomenon, Fujita claimed, had caused at least one commercial jetliner to crash.

As Fujita continued his downburst research despite intense doubt and criticism from his peers, he identified a smaller, more intense form of downburst he named a "microburst." Yet even as Fujita's body of evidence grew, many in both the meteorological and aviation communities remained deeply skeptical.

According to Fujita's theory, a microburst presented a unique deadly hazard. Because of its deceptive nature, pilots inside a microburst did exactly the wrong thing at the worst time.

Connors: Shit.

Connors: Ohhhh shit!

The discovery of the "low precipitation" supercell occurred during Kathleen's first ever storm chase in 1973. Large hail, with very little rain, was falling as the rapidly rotating lowering cloud approached.

This photo, taken by Steve Tegtmeier, captures the wall cloud and funnel cloud that became the Spencer, OK tornado of June 8th, 1974. It was this tornado that we televised live.

The rescue of Delta Flight 191. Courtesy of Ft. Worth Star-Telegram.

The second of two wind-caused derailments in ten days that prompted the Southern Pacific to be the first railroad to use WeatherData's storm warning service. WeatherData is now the primary provider of storm warnings to the railroad industry in Canada, the United States and Mexico.

Photo of the 2009 Aurora, Nebraska tornado. Inset is the hook echo associated with the tornado. Tornado photo by Dick McGowan and hook echo by WeatherTap.

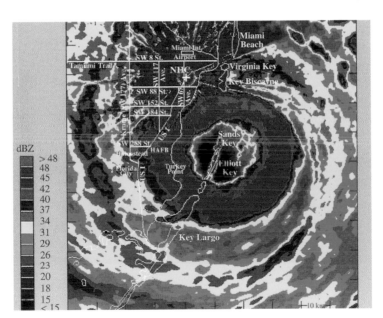

The last image before the failure of the National Weather Service's Miami radar. The red donut-shaped echo is the eyewall of Hurricane Andrew, which produced the most destructive winds. Note that Miami Beach (top center of image), where many reporters spent the night, is well outside of the eyewall. Most of the reporters left town without realizing the havoc unleashed to their south. Courtesy of National Oceanic and Atmospheric Administration.

Tornado-like damage due to Hurricane Andrew's extreme winds. Courtesy of National Hurricane Center

Car floating through lobby of motel in Mississippi as Katrina's storm surge moves ashore. Courtesy of Jim Reed.

Hurricane Katrina damage in Mississippi where a vicious storm surge flattened buildings and carried people out to sea. Courtesy of Jim Reed.

Damaged roof of Superdome. Holes developed in the roof risking structural failure even though winds were less than forecast. Had Category 4 or 5 winds occurred, the results might have been catastrophic for those sheltered inside. Courtesy of FEMA/Jocelyn Augustino.

New Orleans as water pours into Lower 9th Ward. New Orleans flooded after Katrina. Note the maroon-colored barge that drifted into homes and the completely submerged homes behind it. Courtesy of FEMA/Jocelyn Augustino.

Just-rescued couple wondering what comes next. Courtesy of FEMA/Jocelyn Augustino.

Man being carried out on rescue boat as his emotions overwhelm him. Courtesy of FEMA/Jocelyn Augustino.

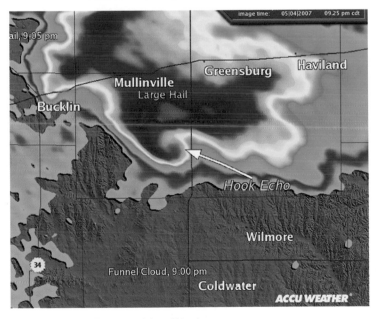

Greensburg hook. Courtesy of AccuWeather.

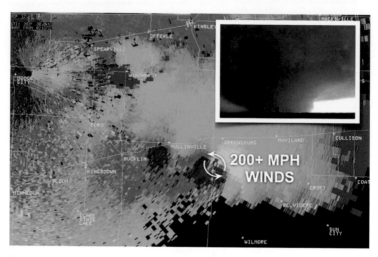

Doppler wind display indicating a wind twist (meteorologists call it a "couplet") of more than 200 mph as the tornado crossed highway 183 southwest of Greensburg just after 9:30pm. The insert photo shows the then one mile wide tornado crossing the highway, illuminated by the near continuous lightning. The electrical poles along the highway fell seconds later. Radar image courtesy of National Weather Service. Inset: Rick Schmidt.

The scene when the sun rose in Greensburg. Courtesy of Larry Schwarm.

Greensburg the morning after the tornado. Courtesy The Wichita Eagle.

Greensburg the morning after the tornado. Courtesy The Wichita Eagle.

Collapsed Greensburg water tower. Courtesy of the Wichita Eagle.

Collapsed Udall water tower. The similarity between the Udall and Greensburg tornadoes extended to the collapsed water towers. In both tornadoes, the support "legs" were collapsed into a shape resembling a child's swing set. Courtesy of Udall Tornado Museum.

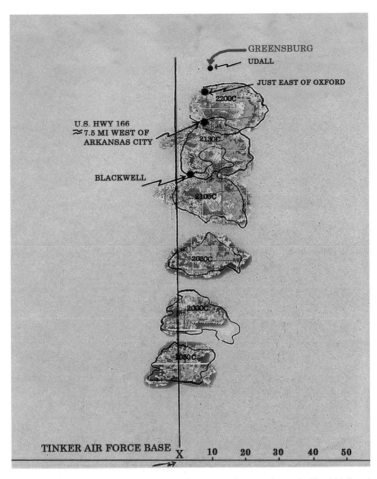

Udall radar tracing with Greensburg echoes superimposed over it. The Udall and Greensburg tornadoes were amazingly similar, even down to the shape and texture of the radar echoes at half hour intervals. Artist: Trina Sanders.

Greensburg's "green," energy-efficient recovery has received national attention. Homes, businesses and a new arts center can be seen in the town as it recovers from the tornado. Photos by Katherine Bay, composite by Karen Ryno.

Amongst the rubble, Tyler McIntosh found a new friend. Greensburg survived because its people survived thanks to the warnings. Courtesy of the Wichita Eagle.

The final sound on the CVR of Delta 191 is a three-second crunching, crashing sound that, when played back, seems to last much longer than the actual three seconds. It is followed by abrupt silence.

The microburst, a phenomenon that many meteorologists said did not exist, had claimed another commercial jetliner and 137 lives.

* * *

Careful analysis of the words and tone of voice on the CVR indicated that both First Officer Price and Second Officer Nassick were concerned about the rain shower. They pointed out the lightning to the captain about ninety seconds before the aircraft touched the ground for the first of four times, starting a full mile short of the runway. After the captain brushed off the comment about the lightning, Nassick remarked sarcastically about Price getting to fly the "good" legs.

In Delta's crew culture, the captain was in firm control, even though at times this particular captain appeared not to be paying full attention ("What?" "Where?"). And until he shouted "toga," seven seconds before first contact with the ground, Captain Connors was clearly committed to the landing. Though the cockpit voice recorder records the sound of engines straining to deliver maximum thrust, by the time the pilots "pushed it (the throttles) way up" to try to gain airspeed and altitude, the battle had already been lost.

One-third of all microbursts are not survivable.

To a Delta 191 passenger in the windowless middle seats of the two-aisle airplane, it probably seemed as if they had safely landed. Actually, the plane touched down in an empty field north of the airport. It bounced (as planes often do on the runway when landing) and touched the ground a second time. Delta 191 then gained altitude for a second, but as it crossed Texas Highway 114 north of the airport, one of the engines clipped an automobile (decapitating the driver on his twenty-eighth birthday). Three-point-six seconds elapsed between the first touchdown and contact with the car.

After hitting the car, the plane skidded on the bottom of its fuselage, shedding parts along the way. It veered left of the runway approach and began to disintegrate, accounting for the three seconds of "crashing" sound on the CVR. The aircraft then struck a large water storage tank and exploded. The water from the ruptured tank poured over what was left of the fuselage, creating a river of fire as the burning jet fuel floated to the top. The deep water made escape difficult. The survivors had to fight the flowing water, torrential rains, and hurricane-force winds from the microburst they had just flown through.

While the few survivors were scrambling for their lives, the microburst continued to move slowly south. When it finally reached one of the sets of airport weather instruments on the east runway complex, wind gusts reached 100 miles per hour, while the west side of the airport experienced dry conditions with partial sunshine. There, it never even rained.

The tower controller, Gene Skipworth, after telling Delta to "go around" (i.e., abort the landing, climb, circle the airport, and get back into sequence to land), saw the crash occur. He instructed all aircraft on approach to the airport to go around, and then he hit the button to scramble the crash trucks.

Within seconds, alarms were screaming in the three fire stations at DFW. Forty-five seconds after the wreckage came to rest, three fire trucks were on the scene. Though their efforts were hampered by the rain and extremely high winds, most of the flames were extinguished within ten minutes of the crash.

The first medical units were on scene within five minutes of the crash. Nine minutes after the crash, medical assistance was requested from ambulances and hospitals based outside the airport. However, in the confusion and stormy conditions, some jurisdictions did not understand what was requested and sent fire trucks rather than ambulances.

Captain Mica Calfee, one of the firefighters responding from the community, has written about the experience. He observed his heavy, large open-topped fire truck being moved by the intense wind, and he had to fight to keep from being "blown away." His crew had trouble getting their truck quickly to the accident scene.

Calfee and a fellow firefighter began to walk toward the crash site when they came across bodies lying facedown in the water. The victims had lost their outer clothing and appeared to be uninjured, yet they were deceased. Despite participating in a disaster drill at DFW Airport just a few weeks before, Captain Calfee says he and the other rescuers were completely overwhelmed by what they saw.

About this same time, the television stations in Dallas and Fort Worth were breaking into their six o'clock news with word that some type of aircraft had crashed.

* * *

This is how the *New York Times* of August 13, 1985, covered the National Transportation Safety Board's (NTSB) release of the Delta 191 air-to-ground communications:

> The pilot of Delta Air Lines Flight 191 voiced little concern about a thunderstorm just before the jumbo jet crashed, killing 134 people . . . About two and a half minutes before the crash, controllers announced "variable winds" at the airport's north end, where the Delta jet was to land. Controllers then told the pilot to reduce speed. Experts say speed is crucial to combat wind shifts.

Note the phrase "to combat wind shifts." Neither the word *downburst* nor the word *microburst* appears in this article. This is illustrative of how skeptical the FAA, the NTSB, the aviation community, and many meteorologists were of Fujita's eight-year-old downburst theory.

While Fujita's discoveries regarding mesoscale analysis, tornadoes, and severe storms were gradually (and sometimes grudgingly) being accepted as other meteorologists replicated his findings, his downburst theory was just too much. So many meteorologists had studied thunderstorms in so many research projects yet none had detected a downburst in *their* research. The level of envy-driven skepticism was unprecedented in our field.

From 1964 to 1985, microburst-related commercial airline crashes killed 503 people. Why were there no known commercial airline crashes due to microbursts prior to 1964? Because most commercial airline planes prior to 1964 were propeller-driven aircraft.

While jet travel is usually superior to propeller-driven aircraft, the performance characteristics of jet engines increase the plane's vulnerability to the form of wind shear (a rapid change in wind speed, wind direction, or both) caused by a microburst.

When the pilot of a propeller-driven plane moves the throttles forward, the propellers spin faster almost instantly, quickly lifting the plane. Jet engines take a moment to respond to the throttle movement, and that short interval of time can be the difference between a crash and a safe recovery.

Today, in another triumph of meteorology research, commercial airline crashes due to microbursts are almost unheard of; the last was in 1994. The story of how microbursts were conquered, with the saving of thousands of lives, again centers around the remarkable Ted Fujita.

* * *

On a hot summer day in June 1975, Eastern Air Lines (EA) Flight 66 crashed on final approach to JFK International Airport in New York. As the plane approached the runway, it ran into a violent thunderstorm. In retrospect, there was almost an air of inevitability surrounding the crash—every pilot knew that flying into a thunderstorm was a bad idea in theory, but too many pilots at the time thought they had the "right stuff." Flying into a thunderstorm was something that claimed the lives of *other* pilots.

Reports to the JFK control tower from an aircraft awaiting takeoff that it was being buffeted by the storm's high winds were disregarded by the air traffic controllers and thus not relayed to the aircraft on landing approach. Another airplane landing ahead of EA 66 barely avoided crashing. The flight crew of EA 66 knew there was bad weather ahead—it was visible on their onboard radar—but pressed on anyway. One hundred and twelve people were killed.

The National Transportation Safety Board (NTSB) determined the probable cause of the crash of Eastern Air Lines Flight 66: it was "the aircraft's encounter with adverse winds associated with a very strong thunderstorm located astride the ILS [instrument landing system] localizer course, which resulted in high descent rate into the non-frangible approach light towers. The flight crew's delayed recognition and correction of the high descent rate were probably associated with their reliance on visual cues rather than on flight instrument reference. However, the adverse winds might have been too severe for a successful approach and landing even had they relied upon and responded rapidly to the indications of the flight instruments."

To most people in aviation, the NTSB is the final word, and the NTSB believed the cause of the accident to be "adverse winds." And it was—sort of.

The FAA began work on a low-level wind-shear alert system (LLWAS) based, in part, on the NTSB's findings. LLWAS is composed of an array of aerovanes placed near runways. An aerovane is a device resembling a miniature propeller airplane sans wings mounted to the top of a pole; the device simultaneously measures both wind speed and direction. The data from each of the aerovanes is sent to a dedicated computer in the airport control tower. The computer searches the data for unusual patterns of wind speed and/or wind direction. If the computer detects out-of-the-ordinary conditions, an alarm goes off in the tower so controllers know to inform pilots.

The scientific consensus in 1976–1977 was that a "gust front" (a small wind shift often associated with rain-cooled air from a thunderstorm; or when cool ocean air flows over hot land) had produced the wind shear that doomed EA 66. The LLWAS system was designed to detect these frontal wind-shear situations.

But to Fujita, who had been called in to investigate the Eastern Air Lines crash, something seemed unusual. Other planes had penetrated the thunderstorm and landed safely, though some had had close calls. Why was EA 66 the aircraft that crashed? What tipped the balance from a near-miss to a crash? Fujita set out to answer this critical question, a question none of the other investigators had even thought to ask. Knowing the right question to ask was part of Fujita's genius.

Fujita conducted a detailed study of the eleven aircraft that landed safely ahead of EA 66. He studied the weather, the radar, and flight paths, and he talked with the surviving crews. He conceived a theory based on his investigation of EA 66 and on some of the unexplained damage patterns he'd observed during his survey of the 1974 record tornado outbreak.

The theory was meteorologically unconventional. In some important ways, it differed from the NTSB and FAA's interpretation of the events at Kennedy Airport. Fujita, along with Horace Byers, published their findings of the Eastern Air Lines crash and theorized the existence of a "downburst"—a rapidly sinking column of air that originated in a thunderstorm and then spread out, and accelerated when it reached the ground. As the air spread out, it could reach speeds of 70 miles per hour or more. A pilot flying through the sinking air, with its rapid change in wind speeds and directions, would be severely challenged to keep control of the plane.

Fujita's theory of the downburst made perfect sense to me, especially since it also explained the intense localized damage sometimes seen during severe thunderstorms in the central United States. Those unexplained damage cases were often called "tornadoes," especially when they occurred at night (when no one would expect to see a funnel cloud unless there was lightning to illuminate it), for lack of a better explanation.

But Fujita's article, even though it was published in a peer-reviewed scientific journal, touched off a firestorm. The headline in *Science News* was "Are Downbursts Just a Lot of Hot Air?" A number of meteorologists believed Fujita had simply misunderstood the nature of a thunderstorm's downdraft. Fujita later wrote in his memoir that the controversy upset him greatly and it caused him "many sleepless nights."

In order to be "scientific," theories and experimental results by definition have to be repeatable and *not able to be made false*. Obviously, if there were downbursts at Kennedy Airport in June 1975, there would be downbursts elsewhere. The goal was now to find them. Meteorological experiments were planned with the goal of detecting and measuring downbursts.

But support for Fujita's theory would come in an unexpected way.

* * *

My college friend and former roommate Steve Amburn was visiting us in Wichita during Independence Day weekend, 1978. It was a hot Saturday with temperatures just above 100 degrees. In the heat, thunderstorms popped up over the Wichita area, so we decided to storm chase.

On U.S. Highway 54, we drove to the west side of the town of Andover. When we saw a thunderstorm building rapidly just to the south, we turned off 54 onto a rural road to get a better view. As we topped a hill, we observed a rain shaft below the developing storm. Nothing unusual about that; the open terrain and lack of haze in the Great Plains often allow storm features to be observed that are difficult to see elsewhere.

But by training our eyes on the rain, it became obvious this was not an ordinary rain shaft; we could *see* it descend! Normally, the descent of rain is so slow it is indistinguishable to the eye.

The rain shaft seemed to slam into the ground and then spread out rapidly, like water poured from a pitcher onto the floor. This pattern was something neither of us had ever seen before.

Then it happened: rain at the edge of the rain shaft "fell" up! Gravity would suggest this was impossible. It then curled back toward the main rain shaft.

The first microburst ever photographed. Raindrops that are blown out of the microburst by high winds then curl upward in the extreme wind shear. Photo copyright 1978, Michael R. Smith.

As the rain shower moved to the east, we changed position to get a better angle, and it was then that we were hit by a blast of 50-mile-per-hour wind that was completely unexpected.

When the strong gust of wind struck us and nearly blew the tripod over, the thought occurred to me: *Have we just been hit by a downburst?* The visible fall of the rain shaft suggested the rain was being accelerated, and the never-before-seen curl of the rain suggested very strong winds were in play.

Steve and I discussed the phenomenon on the way back home and theorized we had witnessed a downburst. I had been taking pictures, and they developed well; so I decided to call Fujita, tell him about our experience, and offer my photographs as proof. I mailed off the photos and he called a few days later. He was beyond excited: I had indeed proved his theory. We began a friendship that lasted until his death, and downbursts passed from theory into fact.

Downbursts were further confirmed by Project NIMROD (Northern Illinois Meteorological Research on Downbursts), conducted in the Chicago area around O'Hare International Airport. In later papers, Fujita defined a "microburst" as a downburst four kilometers (2.5 miles) in diameter or smaller, and a "macroburst" as a larger downburst.

Fujita published several additional scientific papers along with his book, *The Downburst*, but skepticism remained in spite of the growing toll of aircraft-related downburst deaths.

As Fujita's research progressed, it became apparent that downbursts were not just causing crashes on landing (Allegheny 121 at Philadelphia, Royal Jordanian 600 at Doha, Pan American at Pago Pago, Ozark 809 at St. Louis, and others); they were also causing crashes to occur during takeoff (Continental 63 at Tucson, Pan American 759 at New Orleans, and Continental 426 at Denver).

Pan Am's 1982 downburst crash at New Orleans International Airport occurred during takeoff. Strong thunderstorms in the vicinity of the airport were setting off the LLWAS wind-shear alarms (put in place as a result of Eastern Air Lines 66), but because Fujita's theories were not yet accepted, pilots generally viewed them as false alarms. The Boeing 727 began its takeoff down runway 10, accelerating to the east. Because of the alarms, the Pan Am pilots intentionally gained extra airspeed for takeoff, since more airspeed gives more maneuverability. In spite of the extra airspeed, when the 727 entered the microburst, the aircraft immediately sank, hitting trees and a power line. It then crashed into a residential neighborhood.

The plane was loaded with 8,000 gallons of jet fuel and turned the area into an inferno. All 145 people on board were killed, along with eight people on the ground.

The NTSB response to this accident was in some ways contra-dictory. It concluded that wind shear was responsible for the crash and complained that wind-shear forecasting techniques were inad-equate. But this conclusion was misleading: the wind-shear detectors at the New Orleans Airport *had* detected the wind shear and sounded the alarms in plenty of time to delay the takeoff. Because Fujita's microburst theories were not accepted, the NTSB had trouble fully comprehending what had occurred in New Orleans.

Accidents and incidents continued. A United Airlines flight taking off in a 1983 Denver microburst struck a light standard, damaging the aircraft and forcing an emergency landing. President Reagan aboard *Air Force One* was nearly killed when his flight landed moments before a downburst struck Andrews Air Force Base, with wind gusts to 150 mph.

* * *

Almost immediately after the Delta 191 rescue operations were completed at the northeast edge of DFW International, Fujita was summoned. He flew to Dallas and began his unique brand of methodi-cal research. Unlike most meteorological researchers who use graduate students for fieldwork, Fujita himself flew over the airport, surveyed the on-site instrumentation from a cherry picker, photographed the exact location of the anemometer (wind speed) and wind vane (direc-tion), and collected every scrap of data he could. With Fujita, one never knew which type of data might turn out to be crucial.

In addition to the data collected by the instrumentation at the airport, Fujita collected weather satellite imagery, radar data, eyewit-ness reports, and data from the cockpit voice recorder and flight data recorder. Once he had all of the data, he began to weave it into a coherent picture.

Fujita wrote an entire book, *DFW Microburst on August 2, 1985*, on the DFW microburst. He and the other researchers produced a detailed explanation of how and why Delta 191 had crashed.

As the L-1011 neared the north end of the runway, it gradually slowed and descended along what landing aircraft call the glide slope. When the plane initially encountered the microburst, the leading edge of the wind struck the aircraft. And just as I'd observed raindrops "falling up," Delta 191 encountered high winds and rising air. This increased the speed of the aircraft and lifted it above the descending trajectory it was supposed to follow.

This is where the insidious nature of the microburst presents itself. Almost instantly the plane went from being too high to nose-diving toward the ground. As it reached the south half of the microburst, the wind direction shifted from out of the south (a headwind) to the north (a tailwind), causing an instant drop in airspeed and even more sink.

The crew tried to compensate by getting maximum thrust from the engines ("push it way up!"), but it was too little, too late.

* * *

Multiple research investigations, including that of Delta 191, have now conclusively proved the existence of downbursts and microbursts. The result of Fujita's microburst research has been a dramatic improvement in aviation safety. Prior to acceptance of Fujita's theories, microburst-related crashes were occurring approximately every seventeen months. But after integrating Fujita's theory into improved pilot training and warning technology, it was not until 107 months after the Delta accident in Dallas that the next microburst crash occurred, on July 2, 1994, in Charlotte, North Carolina, when U.S. Airways Flight 1016 crashed on landing approach, killing thirty-seven people. As of this writing, we are at 187 months and counting since the Charlotte crash.

* * *

The saga of Delta 191 didn't end when the wreckage was removed.
In some ways, it had only just begun. While investigators were still
combing through the charred wreckage, attorneys in Dallas, site of
the largest annual aviation law convention in the United States, were
already gearing up.

CHAPTER FOURTEEN

DELTA 191: WHY WEREN'T THEY WARNED?

TED FUJITA WASN'T THE ONLY ONE WHO investigated the crash of Delta 191; the National Transportation Safety Board, Delta, and various attorneys immediately began their own investigations. While it quickly became clear that the direct cause of the accident was a level of wind shear (caused by the microburst) that overwhelmed both the pilots and the aircraft, the issue quickly focused on why Delta 191 had been in the thunderstorm to begin with. After all, visibility that day had been nearly unlimited, and just about everyone on the east side of the airport—traffic controllers, pilots preparing to take off, the crew of Delta 191, and the airport weather observer—all had seen the storm.

Why had the crew flown into danger? The more the investigators learned, the more unsettling the question became.

The National Weather Service is charged by law with the responsibility of providing aviation weather data to the Federal Aviation Administration. The FAA, in turn, has responsibility for ensuring that weather information is available to pilots. Seems simple. Unfortunately, both agencies have slightly different interpretations of their missions. And whenever there is overlapping jurisdiction among bureaucracies, trouble often ensues. Such was the case in the early evening of August 2, 1985.

Kathleen Connors, Delta's Captain Connors' widow, sued the Federal Aviation Administration and the National Weather Service for failing to warn her husband of dangerous weather conditions. Delta joined her case. The early betting was on Connors and Delta. In order to understand their position, one has to first understand the convoluted weather radar distribution system that existed in 1985.

In the mid-1980s, the National Weather Service's network of weather radars was a patchwork of twenty-five-year-old WSR-57s and a sprinkling of newer WSR-74Cs. But the term *network* was a misnomer because, for two decades, the NWS had no consistent and *effective* way to get the critical radar picture from the radar to those who wished to use it.

The only data available to meteorologists (outside of the NWS offices that operated an on-site radar) was through a coded teletype report, sent at forty minutes after each hour, such as this: LZK 1133 AREA 4TRW+/+ 22/100 88/170 196/180 220/115 C2425 MT310 AT 162/110.

Even if you were experienced reading this type of report, it was difficult to discern what, exactly, was occurring. These reports were hand plotted each hour in Kansas City to create a national radar chart. This chart was transmitted over the National Weather Facsimile Network (NAFAX) but wasn't received by airports, air traffic controllers, or TV stations until nearly an hour after the original radar observation had been made, making the report useless for storm-warning purposes. An early attempt at creating radar facsimiles was a failure.

As previously discussed, ground clutter was a serious problem with some of the original WSR-57 radar installations, so in a number of cases the radars were moved outside of metropolitan areas. The theory was that by putting the radar fifty miles or so outside of a major city, approaching storms could be better observed by the radar, and the people of the city could be better warned. But there was a major flaw: The meteorologists in the "city" office retained responsibility for issuing the actual warnings, even though they could not directly view the radar. This arrangement would prove fatal, and not just in Dallas.

* * *

The Denver NWS office had warning responsibility for the northern Colorado region, but the radar itself was in the small town of Limon, along Interstate 70 to the southeast. On Saturday, July 31, 1976, intense, nearly stationary thunderstorms developed about sixty-five miles northwest of Denver.

The Limon radar operator grew increasingly alarmed as his radar displayed near-stationary thunderstorms capable of producing tremendous amounts of rain. Unfortunately, he was powerless to warn the public of a potential flash flood. Multiple times he telephoned the Denver meteorologists, who had the warning authority, sharing data and urging them to warn about the possibility of severe flooding. For reasons that are unclear, the Denver NWS office declined to issue a flash-flood warning. The only advance warning of the flood came from private-sector meteorologist John Henz, who recognized the catastrophe in the making, and broadcast his own flash-flood warnings on a Fort Collins radio station.

Simultaneously, there was a reception for visiting meteorologists at the National Center for Atmospheric Research in the Boulder foothills. The scientists were admiring the spectacular cloudscape and brilliant lightning flashes to their north. Some were "toasting" the storms as, unknown to those meteorologists, the persistent storms dropped torrential rains over the watershed of the Big Thompson River.

The banks of the Big Thompson were especially crowded due to the celebration of Colorado's Centennial on that late summer weekend. The Big Thompson, over the eons, had carved out a narrow canyon for much of its length, creating a perfect geographic and meteorological combination to quickly funnel rain into the river.

The thunderstorms ultimately produced more than a foot of rain. Few realized what was occurring until it was too late. A policeman attempting to radio a warning was swept away in his car by a nineteen-foot wall of water. The resulting flash flood, known as the Big Thompson Canyon Flood, killed 145 people.

* * *

In 1985, the Dallas-Fort Worth metroplex had the same split radar/warning responsibility arrangement that existed in Denver in 1976. The NWS radar, as well as the regional weather balloon launch facility, were in the small town of Stephenville, about fifty miles to the southwest of Dallas-Fort Worth International Airport. The data from that radar was fed to two NWS facilities in Fort Worth: the Fort Worth forecast office, located in the federal building downtown and the National Weather Service's Center Weather Service Unit (CWSU) inside the FAA's air route traffic control center near the Dallas-Fort Worth airport. (The Fort Worth office had both an aviation forecast desk and storm warning responsibility for the metroplex.)

Remote radar technology had improved somewhat in the almost ten years since the Big Thompson Flood. Steve Kavouras, an entrepreneur from Minneapolis, received permission from the NWS in the late 1970s to install devices on NWS radars that allowed his customers to "dial in" to receive color radar images over regular telephone lines. Television stations, airlines, commercial meteorologists, and anyone who wanted real-time radar access could purchase one of his receivers to display the data. There was initially only one dial-in connection per radar, so sites with multiple users often got a busy signal.

Because busy signals were unacceptable during threatening weather, customers started installing their own phone lines. Some radars had as many as 120 dedicated phone lines and transmitters, a management challenge for the NWS.

Even with the Kavouras RADAC system and a dedicated line, no one could get a connection during the ten minutes or so from half past the hour to twenty till the next hour, when the radar operator was making his teletype radar observation; or during the twenty minutes or so it took to switch to emergency power during a severe weather situation; or when the operator took the radar off-line to take a special observation during severe weather. So, in addition to the busy signals, there were many times when the radars were unable to create and transmit a RADAC image.

Radar can measure other important variables such as the height of a storm and a storm's structure. But this data was not continuously available to the warning meteorologists, and when it was available, it was through the often illegible wet paper fax or the delayed coded teletype report. These shortcomings would play a major role in the failure to warn of the storm immediately north of Dallas-Fort Worth Runway 17L.

The inadequate radar technology drew the attention of all of the Delta 191 investigators. And as more was learned about how the NWS operated that evening, it looked like they bore a heavy responsibility for the accident:

- The NWS radar technician in Stephenville was at dinner away from the radar console when the microburst-producing thunderstorm developed just north of Dallas-Fort Worth International. Right after he finished eating, he helped launch the evening weather balloon, leaving the radar still unmanned. He did not return to the radar until 6:00 p.m. At 6:04, two minutes before the crash, he telephoned the Fort Worth downtown office to inform it of the storm near the airport.

- The 6:04 p.m. call came too late to allow a warning to have been issued because, according to testimony after the Delta 191 crash, it took the NWS office six to ten minutes to prepare an aviation weather warning for the Dallas-Fort Worth control tower.

- The NWS forecaster in downtown Fort Worth testified that if a level-four radar echo presented itself near Dallas-Fort Worth he would "do nothing with it" (meaning he would not issue a warning).

- The meteorologist manning the NWS aviation desk that evening had no special training in aviation meteorology.

- The wind-shear threat posed by a level-four thunderstorm in a 100-degree atmosphere conducive to microbursts went unrecognized, since Fujita's research was not then generally accepted by the NWS or the wider meteorological community.

What with the ineffective radar technology and sloppy actions of the NWS meteorologists on duty that day, it did, indeed, seem the NWS was at fault. And this was despite the existence of the NWS Center Weather Service Unit (CWSU), a second portal for communication—a fail-safe—with the FAA. Established after the 1977 thunderstorm-related crash of Southern Airways Flight 242 in Georgia, the Center Weather Service Units were manned by specially trained NWS meteorologists and located in the FAA's air traffic control centers across the United States. Their responsibility was, in part, to provide meteorological advice to airport facilities such as control towers, in the form of personal briefings and aviation-related forecasts. One of their Meteorological Impact Statements, a forecast tailored for air traffic controllers and other aviation interests, *could* have been issued for an airport for thunderstorms and wind shear, and would have been received in the Dallas-Fort Worth tower.

The CWSU meteorologist on duty that August day testified to the National Transportation Safety Board that he would not necessarily issue a warning for an ordinary thunderstorm in the vicinity of the Dallas-Fort Worth airport but would issue one if the thunderstorm were to produce wind shear. Unfortunately, no one would ever know if he would have issued a warning in *this* case of severe wind shear because he, too, was at dinner from 5:25 p.m. until 6:10 p.m., four minutes after the crash.

There was one other avenue that might have indirectly led to a warning for Delta 191. The NWS issues severe thunderstorm warnings for the public (as opposed to aviation interests) when it expects winds to gust to 50 knots (58 miles per hour) or more.

The Dallas-Fort Worth microburst produced wind gusts measured at 100 miles per hour. Had the NWS public forecaster in downtown Fort Worth recognized the threat, he could have issued a severe thunderstorm warning for eastern Tarrant and western Dallas Counties. Because the public forecaster was also the supervisory forecaster that evening, he could have *required* the NWS aviation forecaster to issue a warning for the airport. But the NWS public forecaster, too, failed to recognize the signs.

A public severe thunderstorm warning *should* have been issued by the NWS downtown office. The aviation terminal forecast for Dallas-Fort Worth *should* have been amended by the NWS aviation desk downtown. An aviation warning *should* have been given to the DFW tower by the downtown office. And a Meteorological Impact Statement (calling for thunderstorms and possibly wind shear), at minimum, *should* have been issued for Dallas-Fort Worth International Airport by the CWSU.

None of this occurred.

The NWS public forecaster, who should have issued the severe thunderstorm warning and coordinated an aviation warning, was at his post and was trained to issue that type of warning. Why wasn't one issued?

It's likely that it was because of the split radar/warning responsibility arrangement: the radar was in Stephenville, so the public forecaster in Fort Worth couldn't see the texture of the radar echo.

In 1985, it was known that storms that were almost all "core" (as this one was), with only a thin layer of light echoes surrounding the intense, bright red core, often caused strong winds. The storm that day also towered high into the atmosphere for a storm that small, which increased the chance that it would cause strong winds.

That the texture and height indicated a dangerous storm was demonstrated by the fact that the Stephenville radar technician immediately called the downtown Fort Worth office upon returning from dinner. The rapidly growing height of the thunderstorm was discernible by the radar in Stephenville but not by the technology available in Fort Worth. The public forecaster in Fort Worth was relying on the Stephenville radar operator to call him with that type of information—but the radar operator was at dinner.

In NWS offices with colocated radar and warning responsibility, forecasters would have known the Stephenville radar operator was at dinner and thus would have walked to the on-site radar screen to view the dangerous storm for themselves. But this was not possible in the split-office situation, so the public forecaster who was in a position to issue a warning didn't have the data he needed.

* * *

But even if all of the NWS employees had been present and paying attention, there was still no real mechanism in place to *instantly* convey a threat directly to aircraft. This faulty system is one that continues, to some extent, even today.

If a NWS severe thunderstorm warning had been issued by the Fort Worth office to the public, it would have been instantly broadcast by TV and radio stations in the Dallas-Fort Worth area. The warning also would have been placed automatically on NOAA weather radio, ensuring that local highway patrol and law enforcement received it. About the only people who would *not* have received the warning would have been those in the DFW air traffic control tower and pilots in the aircraft on approach to the airport! This ironic situation was due to the Federal Aviation Administration's elite attitude toward weather observations. If it doesn't originate as an aviation weather product, it does not go into the aviation weather system.

FAA weather observations are more detailed than the observations shared with the public. For example, instead of "mostly cloudy," the aviation observation will report that "three-quarters of the sky is covered with clouds, with cloud bases 3,000 feet above the ground." As you can imagine, accurate, detailed information of this nature is critical when airplanes are landing and taking off.

In most cities, the NWS office was located at the airport. The Dallas-Fort Worth Airport was an exception. It had a NWS contract observer—a private-sector employee certified by the NWS to record the official airport weather observations to be used by the aviation community and the media. In the 1980s, the airport weather observations (cloud cover, visibility, winds, temperature, etc.) were done completely by human beings in much the same way as Joe Audsley had recorded the weather the evening of the Ruskin Heights tornado.

Were there no heroes in the Dallas aviation weather community that evening? Actually, there was one, Salvador de Prete, the NWS contract weather observer at DFW Airport.

As Delta 191 was approaching the airport, de Prete took the routine hourly observation, as required, at 5:51. He recorded scattered (less than half the sky) clouds with bases at 6,000 feet, visibility of 11 miles, temperature of 101°F, temperature of the dew point at 65° (a measure of the humidity), and winds from the east-southeast at 8 knots (10 miles per hour). These observations were accurate and representative of the weather conditions at de Prete's observation location, the Delta maintenance hanger adjacent to runway complex 17L. (However, the wind instruments were on the east side of the east runways, to the east of American Airlines' Terminal C.) Though de Prete's 5:51 p.m. observations don't seem to indicate severe weather, he was becoming increasingly concerned about a small thunderstorm visible to the north.

*Micro*bursts, by their very nature, are small, 2.5 miles or less in diameter. The growing thunderstorm just north of the airport could be seen from de Prete's observation post and was noted in his observation in a section called Remarks.

Because of his concern about the approaching thunderstorm, the moment de Prete finished recording and transmitting the hourly data, he ran back outside and saw that the weather was, indeed, getting worse on the north side of the airport. He rushed back into the hanger, sprinted up two flights of stairs, recorded a special observation and transmitted it, and then ran back outside again.

That observation, completed at 6:05 p.m., recorded that clouds were still scattered at 6,000 feet, the visibility was 10 miles, the winds were from the east-northeast at eight knots, *and* there was a thunderstorm in progress (*thunderstorm* was defined as the observer actually hearing thunder) at the airport that began at 6:02 p.m.; the thunderstorm extended north through northeast of the airport, moving slowly south. There was occasional lightning and falling rain visible to the north and northeast. De Prete's observation was transmitted at 6:07 p.m., exactly one minute after the Delta 191 crash.

Even though de Prete, the contract observer, had done his job flawlessly, his observations could not fully describe the perilous situation. Dallas-Fort Worth Airport is huge—larger than the island of Manhattan. The wind instruments at the center of the airport were recording nothing out of the ordinary as hellish winds were tossing the Delta L-1011 around just two miles to the north.

From an area of the airport south of the storm, the rain shaft containing the microburst appeared as a thick, gray-silver opaque mass that looked even darker and more menacing than usual because of the high cloud base. One pilot, viewing the mass of rain while waiting to take off, thought it was a tornado. He began to speculate what to do if a tornado approached his grounded commercial jet.

The Dallas-Fort Worth Airport control tower was also located south of the storm. The six air traffic controllers working in the glassed-in tower cab could all clearly see the storm and Delta 191. The passengers' attorneys contended, at trial, that the controllers should have warned the plane. However, the controllers, while they did have a measure of weather training, were not meteorologists. Their job was to pass along the official weather information.

The controller handling the arrivals for the east runways at Dallas-Fort Worth saw the storm but, as procedure dictated, gave the *official* weather observations, which did not include the "remarks" about the thunderstorm to the north. And until a few minutes after the crash, when the tower received de Prete's "special" observation, there was still no *official* weather observation of a thunderstorm *at* the airport.

The rain shaft associated with the microburst was so opaque the controller could not see Delta 191 enter its north side. When the plane emerged on the south side, the controller immediately recognized that the aircraft was not in the proper configuration for landing. Following procedure, he told Delta 191 to "go around": abort the landing attempt, gain altitude, and get back in line to land. A few seconds earlier, Captain Connors had had the same realization and shouted, "TOGA!" But it was too late. As the controller was speaking, DL 191 slid along the ground and burst into flames.

While the National Transportation Safety Bureau, the Federal Aviation Administration, and the attorneys representing Kathleen Connors, the passengers, the crew of Delta 191, Delta itself, and other parties were finding the NWS's performance to be subpar, Captain Connors' performance itself came under scrutiny.

The Delta 191 black box transcript and voice recording indicated a captain who seemed to be disengaged from the landing process (indicated by his questions "What?" and "Where?"). When his crew pointed out that they were flying into a thunderstorm and caution seemed to be advised ("Lightning coming out of that one." "Where?"

"Right ahead of us"), Connors was telling the tower controller that they were "out in the rain and it feels good."

Landing in a thunderstorm was clearly against both FAA and Delta Airlines' rules. The thunderstorm could be clearly seen and was mentioned in the cockpit while there was still plenty of time to abort the approach. Why would an experienced flight crew fly right into it?

The positions rapidly became etched in stone. Delta: The plane crashed because the government (NWS) failed to tell Connors about the storm. The United States and the NWS: Connors and his crew had all of the information they needed to abort the landing. This would be the key conflict at the trial.

CHAPTER FIFTEEN

THE DELTA TRIAL

WHEN GIANTS SUCH AS THE GOVERNMENT OF THE United States and Delta Airlines clash in court, it's morbidly fascinating—like watching two scorpions engaged in to-the-death combat.

The government will typically draw things out, making the process as painful as possible for the plaintiffs' attorneys, in the hopes that the attorneys will give up or have their resources depleted. In cases of mass casualties, it's not uncommon for a plaintiff's attorney to invest six-figure sums before a trial even starts.

The Delta 191 case, however, pitted some of Texas's richest attorneys (as well as attorneys from other states) against the government. These individuals had considerable personal and business resources they could pool. They were not about to back down and go away. Their strategy was to get maximum publicity and sympathy for their clients, the families of the crash victims, in order to obtain the largest possible judgment or settlement.

When the trial finally opened, the lead plaintiff was Kathleen Connors, who claimed that the government's "failure to warn" caused the death of her husband. Other crew and passenger cases dovetailed with Connors'. Damages asked were between $150 million and $200 million (in 1986 dollars).

In disaster situations, such as airline crashes, where there are many plaintiffs to which the same facts apply, the cases are usually consolidated, and the judge appoints a Plaintiff's Steering Committee, a group of attorneys with special expertise to manage the case to determine the facts and liability. If liability is found, separate hearings are held to determine the damages to be awarded to each individual plaintiff. Such was the case with *Connors v. the United States of America*.

While each side in the legal system is entitled to a competent advocate, at times it seemed like the egos of the attorneys were far more important than getting to the heart of the matter. The bench trial (a trial in front of a judge, not a jury) lasted an unbelievable fourteen months, whereas in general, bench trials are shorter than jury trials because there is less grandstanding.

Here is a taste of the absurdity of this trial: Attorney John Herrick of Fort Worth, who represented one of the individual plaintiffs, begins the cross-examination of a government expert, Joseph Beaudoin, who has testified that the six people manning the DFW control tower did not observe any weather conditions that were a threat to Delta 191:

> **Q.** Do you agree with me, Mr. Beaudoin, that heavy rain is more hazardous to flight navigation that [sic] light rain?
>
> **A.** I wouldn't agree with that . . . Heavy rain is not a detriment to flight.
>
> **Q.** Would you agree that a thunderstorm is more hazardous to flight navigation than light rain? Would you agree?
>
> **A.** I would agree that generally a thunderstorm could be more hazardous than any rain area . . .

Q. Now with reference to the six FAA employees in the DFW tower. I'm going to read to you a poem called "The Blind Men and the Elephant." Very short. It's by John Godfrey Sachs.

It was six men of Hindustan,

To learning much inclined,

Who went to see the elephant,

Though all of them were blind.

That each observation might

Satisfy his mind.

The first approached the elephant

And happening to fall

Against his broad and sturdy side

At once began to bawl

"Bless me but the elephant

Is very like a wall."

The other blind men, of course, felt different parts of the elephant—and described it as *spear* (the tusk), *a snake* (the trunk), *a tree* (the knee), *a fan* (the ear), and *a rope* (the tail).

And so these men of Hindustan

Disputed loud and long.

Each in his own opinion,

Exceeding stiff and strong.

Though each was partly in the right

And all were in the wrong. [emphasis Herrick's]

[Mr. Herrick, continuing] Now I'm going to ask you, Mr. Beaudoin . . . would you not agree that those six men in that tower who did not see the hazardous [weather] condition were like the six blind men of Hindustan? Would you agree?

A. Obviously I wouldn't agree . . . maybe the blind men were in the Delta 191 cockpit because they saw lightning, they saw weather, and they continued their approach.

Q. I'd like to ask you to substitute an elephant for the thunderstorm for the purposes of this question. Is it not true that if a reasonably prudent controller saw an elephant right north of the runway, the controller could not say [to the pilot] "I see an elephant," [but would have to say "I see what appears to be a fan," or a snake, and so forth]?

A. Sir, we don't cover elephants in our handbook. We talk about elements of weather. We would pass on the elements that we observe (such as lightening [sic], rain, or a thunderstorm) to a pilot and then the pilot can determine whether or not it's an elephant . . .

* * *

Q. [Mr. Herrick, continuing somewhat later] Based on what you know, is it your opinion that [the Delta 191 captain] was of the opinion that he was in light rain when he reported the rain to the controller?

A. That's very difficult to answer because he says he's in the rain, feels good. To me that means that he is encountering rain of some intensity and he is not encountering any turbulence . . .

Q. I'm going to ask you a hypothetical question, Mr. Beaudoin. I am going to ask you to assume that you're a pretty good tap dancer. I'm going to ask you to assume that you enjoy tap dancing. Now assuming those two facts, would you rather tap dance with Gene Kelly in a light rain or in a thunderstorm?

A. Am I inside a building?

Q. No, you're outside.

A. That's my only choice, Gene Kelly?

Q. Yes, sir.

A. [There are] other people I'd rather tap dance with.

Q. Well, I'm asking you to assume that your most favorite favorite person to tap dance with would be Gene Kelly . . . would you rather tap dance with Gene Kelly in the light rain or in the thunderstorm?

A. I probably would like light rain. We could go with "Singing in the Rain" and dancing in the rain, things like that.

Q. Because it would feel good, wouldn't it?

A. Probably wouldn't feel too good if it's raining on your . . . I wouldn't want to tap dance in the rain, period, but if I had to pick light rather than heavy, I guess I would pick light.

And so it went, on and on like this, for months.

A key player for the U.S. Department of Justice was defense attorney Kathlynn Fadely, a former pilot who first became a trial attorney for the aviation section of the Department of Justice in 1979. A major point of discrepancy in the trial was whether the plane had flown into a simple rainstorm or a thunderstorm with lightning and thunder. Fadely contended that the storm was not an official thunderstorm. But her defense was based on much more than semantics. Fadely's scientific background (she had a bachelor of science degree from the University of Georgia and was a licensed pilot) allowed her to create an entirely new courtroom weapon: a groundbreaking video reconstruction of the crash.

This was the first computer simulation ever to be used in a major trial. The second-by-second crash reconstruction was on a videodisc, which produced a clearer picture than that produced by a videotape. Further, videodiscs could stop instantly, play in slow motion, and show an event both forward and backward.

In order to construct the meteorological elements of the simulation, the government's team relied heavily on Fujita's reconstruction of the weather surrounding the accident. The team pieced the mountains of data that Fujita had gathered using his unique methods together with the audio obtained from the black box cockpit voice recorder, and the data from the digital flight data recorder (DFDR), the second black box on all commercial aircraft. Remember that the DFDR recorded the altitude, the position of the aircraft, the position of the aircraft's controls, and other similar data. Finally, data from expert witnesses hired by the government was added to the mix. After each of these layers of information had been added, a very complete view of the flight, second by second, could be obtained by playing the disc.

* * *

Expert witnesses were utilized by the attorneys for the plaintiffs and defendants alike. In a major trial such as this, being an expert witness is a great challenge. You must first be a top-of-the-field scientist, but you must also be able to explain scientific concepts to the members of the jury (or in this case, to a judge) in a way they can understand.

The experts in Delta 191–related litigation had the following challenges:

- Explain the inner workings of thunderstorms

- Explain how some thunderstorms can product microbursts

- Explain what a microburst is

- Explain why this thunderstorm caused a microburst and why they were confident in that belief

- Explain why this microburst affected the Lockheed L-1011 when it did not affect the aircraft that landed immediately ahead of Delta 191

- Explain how the aviation weather warning system was supposed to work and how it worked that day

As the expert gathers data, begins a detailed analysis, and then tests that analysis from varied points of view, he or she must at the same time ignore what the press and others might be speculating about the case.

After an exhausting, emotionally draining, groundbreaking, and sometimes ridiculous fourteen-month trial, Judge David Belew ruled in favor of the United States (the National Weather Service and Federal Aviation Administration). The judge found that while the government's (NWS) performance had been poor, ultimately Captain Connors had all the information he needed to determine he was flying into a thunderstorm, and he should not have continued the landing.

Judge Belew specifically cited the reconstruction simulation of the flight as contributing to his finding of liability.

Delta had lost its case. The plaintiffs' attorneys issued press releases. The families of the victims received their compensation from Delta and its insurance company rather than from the taxpayers.

* * *

The fourteen-month trial surrounding the crash of Delta 191 was a turning point in meteorology, aviation, and in aviation meteorology, in particular. For Delta 191 substantiated Fujita's microburst theory. Never had so many meteorologists reversed their skeptical opinions so quickly.

Because of the trial, the FAA changed its regulations regarding thunderstorms and wind shear. The NWS, recognizing that the split radar/warning responsibility contributed to some of the worst failures in its history, changed its entire forecast structure in the 1990s so that radar data and warning responsibility were colocated 100 percent of the time.

Much follow-on work based around Fujita's research began in order to train pilots how to avoid microbursts, and if they were to inadvertently fly into microburst wind shear, how to escape it (if possible). John McCarthy, my former atmospheric physics professor from Oklahoma University, was instrumental in convincing the FAA to deploy Terminal Doppler Weather Radar (TDWR) to detect microbursts and other hazards in time to warn pilots.

Another such person to expand Fujita's research was Patrick Clyne, a senior captain for Northwest Airlines. Northwest had been an early leader in aviation meteorology. In 1966 it created its Turbulence Plot (TP) system that allowed pilots to plot and visualize hazardous conditions while in flight. No other airline had anything like it, and eventually, Northwest sold its TPs to other airlines.

Clyne had worked with the Air Line Pilots Association (ALPA) to convince the NWS to abandon a proposal to make its weather radar data for external users less detailed. As wind shear began claiming more jets, Clyne became more interested in weather. So he began devoting more of his time to weather and turbulence issues. Concurrently, he started working to heighten awareness of weather hazards in the pilot community.

Clyne remembers the controversy as to whether there "really were microbursts." Within the pilot community, there were "differing theories about how wind shear worked. There was controversy about what weather conditions occurred near the ground when one of these accidents took place."

But since the microburst doubters had thrown in their towels after Fujita's analysis of the DFW microburst, the job now was to figure out how to stop downburst accidents. Terminal Doppler Weather Radar would serve to detect most microbursts and allow air traffic controllers to warn pilots. But not all airports had TDWR, and a small number of microbursts cannot be detected by TDWR.

Clyne, McCarthy, and other scientists were part of an effort to compile the "Wind-Shear Training Aid," first published in February, 1987. The Federal Aviation Administration brought in Boeing, McDonnell Douglas, Lockheed, and other aviation companies and contractors to quickly transfer the new knowledge to pilots and airline managements.

Avoidance of the microbursts was the key. After Fujita and others completed the analysis of Delta 191 and other microbursts, the results were fed into flight simulators; researchers quickly learned that about one-third of all microbursts are simply not survivable—once a plane is inside one of those microbursts, a crash is inevitable. And there is no way to tell in advance which microbursts are survivable and which are not.

However, if a flight crew found themselves inside a microburst, the "Wind-Shear Training Aid" provided the best possible escape procedure, crafted through the use of flight simulators and from test flights into storms with specially equipped aircraft.

Did the new knowledge and procedures work? Since the crash of Delta 191 in 1985, there has been only one microburst-related crash of a U.S. airliner—the July 2, 1994, crash of U.S. Airways Flight 1016 at Charlotte, North Carolina, which killed thirty-seven people. And even that crash could have been avoided—if the flight crew had followed the avoidance procedures set out in the downburst training course.

Given the ever-increasing number of people and planes in the air, the number of lives saved due to Fujita's pioneering research that eventually led to implementation of microburst avoidance procedures in the United States is well over two thousand, not to mention the hundreds of millions of dollars of aircraft losses prevented.

In January 1988, two and a half years after the crash of Delta 191, Ted Fujita was awarded the American Meteorological Society's (AMS) Award for Outstanding Contribution to Applied Meteorology. This is equivalent to an Academy Award in meteorology. The citation accompanying the award stated that it was given for "pioneering studies of damaging storms on the mesoscale."

The award was richly deserved and long overdue.

But Fujita's work was not yet finished. He researched Hurricane Andrew in 1992 and identified, for the first time, small-scale vortices of high wind in the eye wall of the hurricane that had caused especially intense damage.

Fujita passed away in 1998. The AMS, at its annual meeting a few months after his death, conducted a special seminar reviewing and honoring his work. It was chaired by one of his former students, Dr. Roger Wakimoto, and was a tremendous success. Fujita's legacy had, at long last, been fully recognized by his peers.

Fujita's microburst work is now routinely used by meteorologists and others throughout aviation. As I was writing this chapter, the following notification of flight delays in Denver was posted on the FAA's air traffic control website:

DELAY INFO				
ARPT	AD	DD	TIME	REASON
CLT		+60	1945	WX:TSTMS
DEN		+15	1943	WX:MICROBURST ACTIVITY
EWR		+15	1934	TM Initiatives:GS:WX:SWAP
IAD		+105	1940	TM Initiatives:ESP:VOL:EnRoute
LGA		+30	1852	TM Initiatives:GS:WX:SWAP
ORD		+15	1823	VOL:Multi-taxi
TEB		+45	1937	TM Initiatives:MIT/MINIT:SWAP

Table above explains that flights were being delayed by 15 minutes in Denver (DEN) and that those delays were increasing in duration ("+") due to microbursts in the area.

Today, microbursts are detected by a combination of sophisticated Doppler radar placed near major airports and an array of wind measuring devices around both major and medium-size airports with commercial flights. Some aircraft even have turbulence detection radar on board. We are far, far safer from microbursts than we were in 1985.

* * *

While the crash and trial of Delta 191 changed the course of aviation safety for the better, the environment is still not what it should be. Traveling by plane is by far a much safer means of transportation than traveling by car. But it could be safer, still, were it not for bureaucracy.

In Chapter 2, I mention the bureaucratic myopia of the Federal Aviation Administration that sees a line between "aviation weather" and "weather." Because severe thunderstorm warnings (and tornado warnings, for that matter) do not *originate* as "aviation products," they were—and are—not allowed to be given to controllers or pilots. This puts airline passengers at unnecessary risk, as the system for distributing public storm warnings is faster and more finely honed than the system for distributing aviation warnings. On Christmas day 2006, a tornado passed across the northwest part of the Daytona Beach International Airport (DAB), causing $33 million in damage and injuring sixteen people in the area. Embry-Riddle Aeronautical University, adjacent to the airport, lost four buildings and a number of planes.

A tornado warning had been issued by the NWS twenty minutes in advance of the tornado reaching the Daytona Beach Airport. Yet tower personnel had no way to receive the warning and were unaware of its existence. A Comair (Delta Express) regional jet from New York's LaGuardia Airport, on approach to DAB, was also unaware a tornado was approaching and, according to newspaper reports, was on a collision course with the storm. The Comair flight was scheduled to land at 1:39 in the afternoon.

Power at the airport went out at 1:40 p.m., and the Comair aircraft was redirected away—not because of knowledge of the tornado, but due to the tornado-related power failure. The tornado hit the airport at 1:45 p.m. (when the Comair flight would have been taxiing), causing the unevacuated tower to shudder. At 2:06 p.m., the Comair flight was able to land safely. When the pilot saw fifty damaged planes and the building roofs bearing holes at Embry-Riddle, he radioed the air traffic controllers, "What happened?" A well-warned tornado was in the area—and neither the controllers nor the pilots knew about it.

The Daytona Beach incident was argued and discussed in aviation blogs, the local newspaper, and *USA Today*. Months later, the governmental finger-pointing continued. The FAA had banned "distracting" weather radios, which broadcast tornado warnings, from control towers; the National Air Traffic Controllers Association countered that weather radios were needed in the towers because tornado warnings are not broadcast on the air traffic communications network; the air route traffic control center said it was the job of the tower to warn the plane; the airport authority deflected attention back to the FAA ("we don't get involved"); and the NWS insisted how valuable weather radios are. No one was taking responsibility and solving the problem.

My opinion: It doesn't matter that tornado and severe thunderstorm warnings do not originate as aviation products. They should be provided to air traffic control towers and to pilots to keep planes out of critical danger.

* * *

The radar system dedicated to detecting aviation-related hazards within fifty miles of airports is called Terminal Doppler Weather Radar (TDWR), though you've probably never seen it in a TV weathercast. For fifteen years, the data these radars produced was hoarded by the FAA, which would not share it—even with the NWS for the first ten of those years. These radars, usually located near major cities, surveyed the atmosphere every minute, as opposed to every five minutes by the NEXRADs, and could have made a significant contribution to the overall NWS warning process.

These days, TDWR *is* shared with the NWS, and finally, after more than a decade and a half, the NWS and the FAA have begun to make part of this data available to the larger meteorological community. In the words of TDWR creator, John McCarthy, "This move to get TDWR data out to a broad array of users has been in the works for a very long time. It's way past time for this, as these data will be very useful. Unfortunately, all too often, the FAA moves in slow and mysterious ways."

In August 2009, as many as eight TDWRs were down for maintenance, including the radars at JFK International Airport and Louis Armstrong New Orleans International Airport—sites of former downburst-related plane crashes—at the height of downburst season. At least some of these eight radars were down for routine maintenance, planned in advance. This routine maintenance easily could have been scheduled in winter, outside of microburst season. Sometimes the FAA seems to forget what its mission is.

While meteorology as an applied science has made tremendous strides in the past fifty years, we have not yet discovered a cure for bureaucracy.

CHAPTER SIXTEEN

WEATHERDATA

I GREW UP IN AN ENTREPRENEURIAL FAMILY: MY father, grandfather, and uncle all started businesses. So, even as a pre-teenager, I knew I wanted to start a weather company someday. That's why I was initially attracted to working in television. I thought I could learn more about business working in the private sector than working for the National Weather Service (NWS).

WeatherData opened for business on August 31, 1981, in Wichita, Kansas, with three employees: Dennis Smith, a high school friend who, like me, had become interested in weather as a result of the Ruskin Heights tornado; Jon Davies, also a meteorologist; and me. My goal was to create a company that would provide more precise, timely, and tailored information that would be financially valuable to businesses. Because we were linked with the local television station, we had superb equipment, including an advanced radar and every meteorological communications network we could get our hands on.

Starting a business had always been my career goal, but I found it to be more challenging than I'd expected. I hadn't fully considered the implications of not having an "off" switch, since WeatherData operated around the clock, every day of the year. If an operational problem cropped up, for instance, we had to fix it—while continuing to operate. Not an easy task. The vital business necessities of marketing, advertising, corporate taxes, accounting, and so on, were skills that I had to learn for myself; they weren't part of the curriculum for meteorology majors. What did serve me well is my ability to see where technology is headed and to combine that with customer needs.

At first, we were primarily a broadcast weather company with KSNW-TV and the television stations it owned in central and western Kansas, and with radio stations from North Dakota to Texas.

In 1983, the *Wichita Eagle* became our first newspaper client. We provided weather maps and forecasts that their artists and typesetters put into the newspaper. In 1985, we moved one step forward by adding embedded typesetting commands into the text forecast, which saved a great deal of labor at the newspaper.

When Apple created the Macintosh computer and gave it graphic capabilities in 1984, I was immediately intrigued. Could WeatherData create entire weather packages on the computer and ship them to the newspaper? If so, we could use our expertise to produce a better product while saving the newspaper money and allowing them to focus on their core mission, gathering and reporting news.

So I purchased a Mac and hired a Wichita programmer to modify MacDraw (a graphics program included with early Macs) to create graphic patterns suggestive of weather (e.g., a jagged line representing lightning to indicate a forecast of "thunderstorms" on the map). It worked! In addition to the *Wichita Eagle*, in a matter of months we had picked up newspapers in Louisville and Lexington, Kentucky, and in Springfield, Missouri.

In 1987, we attended the newspaper industry's annual trade show in Las Vegas. Having never exhibited at a trade show, we weren't sure what to expect. Turns out, WeatherData—and the services we provided—were a hit. Within weeks, we had signed the *San Francisco Examiner* and *Los Angeles Times*. As technology improved, we were able to provide the first-ever full-color, complete weather package (i.e., 100 percent of the content from WeatherData) to the *Rocky Mountain News* in Denver. Our newspaper business continued to grow, with our supplying weather data to many of the largest newspapers from Providence to Los Angeles.

WeatherData was excelling in the area of media weather, but that business wasn't what I ultimately wanted to do. Technology had by now advanced far enough to allow us to achieve my ultimate business and scientific goal: precision warnings of high-impact weather. While this usually meant storms, it also meant temperature extremes, droughts, and other important conditions. It was now time to begin marketing and selling this new service to potential clients.

* * *

All government weather agencies and commercial weather companies work from a common meteorological database that is primarily comprised of data from the government. Managed by the NWS, this data can include temperature readings at airports, weather balloons, radars, images from weather satellites, and other vital information. There are some private sector data networks, such as the two lightning detection networks in the United States. From this common database, forecasts, storm warnings, river stage forecasts, and other meteorological and hydrological products are created. Some of these products require substantial human input (e.g., tornado and hurricane warnings), and some are completely automated.

It wasn't until the late 1980s and early 1990s that the data could be gathered quickly and in sufficient detail to allow tornado warnings, for example, to be created at a central location such as WeatherData, rather than at a local office near the location of the storm.

* * *

In the fall of 1986, a Southern Pacific train derailed at Durham, Kansas, north of Wichita, due to a washed-out track. WeatherData had warned our viewers and listeners of flash flooding in Marion County (where the washout occurred), but the railroad, with a regional dispatch center in Houston controlling the track in Kansas, had no way of knowing what we were forecasting.

So we approached Southern Pacific and met with company officials in Kansas City. No luck. They believed that the Durham washout was an isolated occurrence.

Southern Pacific had pioneered a new type of railcar, the double stack. These amazing railcars enabled a single train, with just two crew members, to handle as much cargo as 250 trucks while burning far less fuel. The tall cars were economical and environmentally friendly.

But they had one big problem: because of their height, they were highly vulnerable to being blown over when winds exceeded 55 or 60 miles per hour.

In the space of ten days in the summer of 1987, Southern Pacific sustained two weather-related derailments of double-stack trains in Kansas, each costing more than $1 million in damages. In both cases, thunderstorms had generated intense winds that blew the cars over.

On July 5, 1987, just after lunch, the phone rang. It was the Southern Pacific regional office in Kansas City. They had reviewed our proposal from the previous autumn and wanted to start our WeatherData service—that day! We scrambled and made the necessary arrangements to plug them into our network.

Just a few hours later, we observed a thunderstorm developing near the railroad track in Stratford, Texas. The radar showed indications of a hook echo, meaning that a tornado might be forming.

We looked at each other and asked, "Do we really want to do this?" After all, Southern Pacific had been our client for just a few hours. And the NWS had issued neither a tornado watch nor *any* kind of

warning. We didn't want to risk issuing a false alarm and losing credibility with Southern Pacific on our very first day. But we thought
the radar indication of a tornado was pretty strong, so we issued our
first-ever track-specific railroad warning: a tornado warning from
Guymon, Oklahoma, to Dalhart, Texas, faxed to the Southern Pacific
dispatch center.

Later that evening, the phone rang. We learned that two trains
headed for Stratford had been stopped by the dispatcher as a result
of our warning. After the warning expired, the dispatchers allowed
the trains to move forward at restricted speeds in case there was any
debris across the tracks.

When the first of the trains got to Stratford, the crew could clearly
see tornado damage on both sides of the track. Our WeatherData
warning had been accurate. Southern Pacific had only had our service
for three hours, and we'd already saved them several million dollars
and ensured the safety of their crews.

Over the next year, Southern Pacific extended the WeatherData
service to their entire railroad system. Two years later, both the Santa
Fe and the Denver and Rio Grande Western railroads signed up for
our services as well.

These days, WeatherData keeps railroad crews alerted to severe
weather for most of the railroads in the United States, Canada, and
Mexico. On July 8, 2009, we began service to Ferromex, the largest
railroad in Mexico. Fifteen days later, we issued a flash-flood warning for the Ferromex Copper River Canyon route. Ferromex sent out
a hyrailer, a truck with metal railroad wheels, ahead of the train to
scout the track. The hyrailer found water flowing over the tracks. The
train was stopped before it reached the remote Copper Canyon, and
the passengers were put on buses. By the time the train would have
arrived in the canyon, the track was washed out and a derailment,
most likely causing deaths and injuries, would have been a certainty.
According to Ferromex's Al Cisneros in an e-mail the next day, "your
service means the difference between life and death."

But it wasn't just railroads benefiting from the Weather Data services. Toyota, the world leader in logistics, hired WeatherData to ensure that floods, ice storms, and blizzards did not block the flow of parts to their new car factory in Georgetown, Kentucky. One winter we issued an ice-storm warning for Toyota. They were able to procure additional parts ahead of the storm so they could continue assembling cars. According to an article in the *Louisville Courier-Journal*, Toyota was able to keep producing automobiles. But the Ford plant in Louisville, which did not have our service, had to shut down and send their employees home when they ran out of parts. In this case, we saved Toyota a great deal of money.

* * *

For WeatherData to perform its mission of warning its clients, it has had to develop some rather sophisticated technology. But we knew other meteorologists would benefit from these devices. The first is called SelectWarn®, which is used by television stations and emergency management to *precisely* warn people located in the path of a storm.

The second is a handheld device called Storm Hawk®. Storm Hawk uses GPS data to center the display on the user's exact location so radar, lightning, storm warnings, and other critical information can be displayed in a usable manner. The value of the GPS is that a storm spotter, for example, can relate the position of storms, even at night, to her location and stay out of harm's way.

* * *

I have attempted to figure out how many lives have been saved and how many dollars of damage have been prevented due to our WeatherData services. Clearly, the number of lives saved is well into the hundreds and is likely in the thousands. Based on figures our clients have given us, the number of dollars saved is far in excess of $100 million.

It is no wonder that the men and women of WeatherData are so proud of our little company.

CHAPTER SEVENTEEN

AMERICA GETS DOPPLERIZED

DURING THE PERIOD WHEN AMERICA WAS PUTTING a man on the moon and demonstrating its scientific prowess to the world, weather radar technology hardly changed at all. There was some nibbling around the edges, but nothing that greatly improved the quality of tornado, thunderstorm, and hurricane warnings.

No one knew it at the time, but improving the primary storm-warning tool—radar—would be far more challenging than anyone suspected. It was only nine years after President Kennedy had committed America to putting a man on the moon that Neil Armstrong took his first lunar steps. It took double that amount of time to get a network of Doppler radars running in the United States.

By 1980, no one was disputing that a new network of radars was needed. The WSR-57 radars that were the backbone of weather radars in the United States had been designed in the mid-1950s and required vacuum tubes that were no longer manufactured in the United States because no other American device used them. To get replacement tubes, the NWS had to purchase them from a single supplier in the Soviet Union. Imagine: The entire U.S. weather enterprise and the safety of its citizens could have been held hostage in a trade or political dispute.

In addition, prior to NEXRAD, the "NEXt generation RADar," deployed by the government from 1991–1996, there were very few radars west of the continental divide. Major cities, such as San Francisco, Seattle, Salt Lake City, Portland, and Phoenix, had little in the way of radar coverage. And with a mushrooming population in the West and limited water supplies, accurate detection and measurement of storms was of increasing importance.

So new radars were absolutely necessary. The question was whether the replacement radars should be Doppler or conventional.

While we hear the term "Doppler radar" all the time, few people actually understand what the word *Doppler* means. Doppler radar is named for a real person, Christian Doppler, an Austrian mathematician and physicist who discovered the Doppler effect.

A common example of the Doppler effect is the change in pitch of a railroad locomotive's horn as a train approaches and moves away from a listener. The horn's high pitch intensifies as it nears and then transitions to a lower pitch as the train passes and moves away. A similar shift occurs in the radar's energy when the winds are blowing toward or away from the radar. The change in "pitch" of the waves can be measured and converted to wind speed.

If you can measure the wind remotely, you can create more accurate warnings of hurricanes, tornadoes, ordinary high winds, and aircraft turbulence, and you can also infer the size of hail and other important meteorological measurements. You can also better inform people to keep out of harm's way.

Sound simple? It is in concept, but it is actually very, very difficult to achieve from both a physics and an engineering standpoint.

The story of how Doppler radar came to save lives begins in the 1950s.

* * *

After the huge death tolls from tornadoes at Waco, Grand Rapids, and Worcester, the attitude of meteorologists—that increases in tornado fatalities are an inevitable part of population growth—began to change as the political pressure to "do something" about tornadoes intensified.

Ruskin Heights was probably the first real taste of success when it came to the Weather Bureau issuing tornado warnings. Joe Audsley showed it could be done, but he was ad-libbing that evening. To achieve any kind of consistent success, better tools and techniques were needed. The thinking quickly gravitated to Doppler radar, that was used by police to measure the speed of automobiles. It might work to measure wind speeds.

The first meteorological Doppler radar was deployed by the Weather Bureau in Wichita from 1957 to 1959, toward the end of a period of exceptionally violent tornadoes. In the wake of the devastating 1955 Blackwell-Udall tornado, Wichita seemed like a good place to base this radar, for it was likely a tornado would eventually form near enough to be measured.

The Wichita Doppler radar was a modest success in that it measured the winds in the El Dorado, Kansas, tornado of June 10, 1958. That tornado, an F4 or F5, killed fifteen people. However, while the radar had determined that a tornado existed and had measured a 200- mile-per-hour wind speed, it had no way of determining how far away the tornado was, could point in only one direction at a time (rather than continuously rotating in a 360-degree circle), and had a three-centimeter signal wavelength, meaning it had very limited storm-penetrating power. These were huge obstacles to be surmounted if Doppler was ever going to be used nationally as an effective warning tool.

Experimental Doppler radar based in Wichita in 1958. Courtesy National Weather Service.

In 1963, about the same time as the pioneering Doppler work was going on near Wichita, a fatal airliner crash occurred in Maryland, with a death toll of thirty-one. The investigation showed that lightning had struck the fuel tanks of the aircraft, causing them to explode. The Maryland crash and investigation, combined with the tornado terror of the past dozen years, caused the government to create the National Severe Storms Project (NSSP), based in Kansas City. The goal was to better understand thunderstorms and the violent weather they could create. The NSSP moved to Norman, Oklahoma, home of the University of Oklahoma's School of Meteorology, in the 1960s and changed its name to the National Severe Storms Laboratory (NSSL).

A talented group of radar meteorologists was recruited and work began to improve methods of severe storm detection. The National Weather Bureau donated the Wichita Doppler to NSSL. However, the NSSL meteorologists quickly discovered the system was too limited to be of much use, other than having proved the theory that it was possible to detect a tornado using Doppler radar. The radar was analog (rather than digital), so it was nearly impossible to transmit, capture, and store data. The Wichita radar was discarded.

Doppler radar creates substantial amounts of data, which was overwhelming the collection and processing systems of the 1960s. And in order for this system to be used for any type of storm warning, Doppler would have to process information instantly.

In 1969, the NSSL obtained a large-antenna surplus U.S. Air Force ten-centimeter radar with improved rain-penetrating power and began work on what would eventually become the prototype Doppler radar, but there was still a great deal of intricate work to be done.

* * *

After viewing the first of the tornadoes spawned by the Blackwell-Udall storm in 1955, Don Burgess continued to mature in the family home in Stillwater until they moved to Oklahoma City two years later. A self-described "weather geek" in high school, Burgess watched and plotted storms and expected to work for the Weather Bureau when he graduated from college. Burgess' father taught at Oklahoma State University and Burgess planned to attend meteorology school there. However, Oklahoma State dropped its meteorology program, so Burgess set his sights on the University of Oklahoma.

While NSSL was gearing up, Burgess was at University of Oklahoma working on his degree in meteorology. While he was still a student, he began working for NSSL. Burgess later described his NSSL employment as a "gift from President Nixon"; because of a hiring freeze imposed by the president, he was ineligible to work for the Weather Bureau.

As the storm-chase program was coming into its own, NSSL reorganized its Doppler program, calling it the Doppler Radar Group. Manning this group were Burgess, meteorologists Ken Crawford and Roger Brown, and a National Oceanic and Atmospheric Administration Corps (NOAA) officer originally from Kansas City, Les Lemon. Burgess and Lemon quickly bonded due to their mutual interest in tornadoes and severe storms. Lemon was several years older than I, got interested in weather because of the Ruskin Heights tornado, and attended Grandview junior and senior high schools.

Throughout the 1970s, the team of meteorologists and engineers at the NSSL in Norman worked tirelessly to make Doppler radar a practical reality. Burgess and Lemon were central to this effort.

The watershed event in the development of Doppler radar was the Union City, Oklahoma, tornado of 1973. It was the first tornado with a complete data set: chasers on site photographing and documenting the storm, Doppler radar, and special weather balloon launches. Until Union City, it had been theorized that Doppler could provide the data for real-time tornado warnings, but the only positive demonstration had been the limited success of the Wichita Doppler in the late 1950s.

Union City proved that a tornado signature could be detected in the thunderstorm wind field as surveyed by Doppler, even if there was no evidence of a hook echo. By the mid-1970s, the usefulness of Doppler had become obvious. Once the Doppler-tornado warning hypothesis was proved, the task shifted to making Doppler a practical reality.

Lemon broke off from NSSL to return to Kansas City, where he worked in the Techniques Development Unit of the National Severe Storms Forecast Center.

Burgess and Roger Brown remained at the NSSL to plan Joint Doppler Operational Project (JDOP). It's one thing to detect a

tornado in a controlled environment; it's another to be able to repro-
duce those results under the pressure of real-time weather-forecasting
and storm-warning operations.

In this project, meteorologists at the NSSL were in controlled
competition with the Oklahoma City office of the National Weather
Service to validate whether tornadoes could, in an actual operating
environment, be detected more effectively with Doppler radar. The
meteorologists at NSSL operated the experimental Doppler radar
while the Oklahoma City meteorologists were using the late-1950s
WSR-57 radar. The goal of the "bake-off" was to determine whether
Doppler enabled meteorologists to issue better and earlier warnings
of severe storms and thus justify the higher cost of new Doppler (as
opposed to conventional) radars.

Once the results were in, it was apparent that JDOP was a big success. Doppler had proved itself. The final JDOP report was issued in
1979, and the NEXRAD program office, established to manage the
creation of a network of Doppler radars, was opened.

Now that Doppler had fully proved itself, one would have expected
the radars to be rolled out during the early to mid-1980s. And one
company, Enterprise Electronics, had the new radars ready to go by
1983. Unfortunately, scientists were dealing with the federal government, where nothing is ever simple or straightforward.

The Doppler radar program was huge by National Weather Service,
and even Department of Commerce, standards. Never had the National
Weather Service had this many dollars to spend on a single program.
Two other agencies, the Department of Defense and the Federal Aviation Administration, were helping pay for the Doppler program and
would use the resulting radar: this further complicated matters.

The NWS was charged with administering the program, and
frankly, they didn't know how to do it. Plus, the NWS has, for its
entire history, been averse to purchasing "off the shelf" systems—
it prefers to design its own equipment, often without considering
whether the additional cost of a custom approach would be worth

the additional cost in dollars and time. So Enterprise radars were not seriously considered, and the NWS started drawing up plans for custom-built radars.

There were three companies in the initial race to win the Doppler contract: Ford-Westinghouse, Sperry (later renamed Unisys), and Raytheon. The competition to build the new radar system quickly boiled down to Raytheon and Unisys. Raytheon had a successful history of building radars for government uses. Unisys had never produced this type of radar, but it had a secret weapon in Les Lemon, who had gone to work on their NEXRAD program.

Unisys got the contract. Almost immediately, Raytheon objected.

While charges and countercharges were volleyed about in Washington, the NSSL Meteorology Group in Norman was charged with developing the computer programs needed to make the NEXRAD output useful. Don Burgess became chief of the NSSL Forecast Application Group, where techniques were developed for the three agencies funding the NEXRAD project. Each agency had unique requirements, and balancing the needs of the three, given the limited, and comparatively expensive, computing capabilities available in the 1980s, was a difficult task.

As originally designed, the new radar had no system to provide the data to external users such as television stations, emergency managers, private-sector meteorologists, and others. This meant the radar that ordinary citizens' tax dollars funded would not be seen on television weathercasts or provided to airlines. When the private sector weather community received this astonishing news at a January 1979 meeting, it immediately pressed the NWS for an explanation. The response? "We didn't think anyone would be interested." Rarely had anyone accused the NWS of that era of being overly focused on the requirements of its "customers."

Faced with a huge outcry at the news, the private-sector weather community worked with the NWS to get a connection to the radar. The NWS, which had been managing more than 100 separate connections to its old WSR-57 radars in highly populated areas such as Chicago, didn't want to deal with this level of external-user complexity with the new radar system.

The obvious solution would have been to contract with a private-sector company or companies paid by the NWS to maintain a communications infrastructure for the new radars that could be used by all units of government, researchers, and the private-sector weather community. Anyone who wanted to would have been able to tap into this data for just the cost of communications.

Again, this being the federal government, the obvious was not what was done. The NWS tried to make the radars' data a profit center. It started a competition to allow companies to bid on up to four *exclusive* connections to the radars, and made the winning companies pay the NWS millions of dollars in user fees for the privilege of spending millions more per year in long-distance connection fees to receive inferior quality data speed and data than what the government was providing to itself!

The result? Any taxpayer who wanted to use this data had to pay for it twice. One payment on April 15 when he or she filed income taxes, and another to a company to actually furnish the radar data the individual had already paid for.

Four companies ended up getting the contracts: Unisys (which built the radar), Weather Services International, Alden, and Kavouras.

The wider meteorological community was justifiably outraged by this turn of events and made insinuations that private-sector weather companies, such as WeatherData, were behind this nefarious plan by making the data so expensive it would block out the little guy, whoever that was. Actually, we were the people arguing most loudly against the plan because it increased our costs tremendously and threatened to restrict our ability to use federally funded data to serve our customers. While it took years, eventually this issue was resolved and the radars' data became widely available at little or reasonable cost.

Meanwhile, the future of the radar itself was in serious doubt.

The NEXRAD radar was dubbed the WSR-88 (Weather Surveillance Radar, deployed in 1988). Unfortunately, the radar was not deployed in 1988; in fact, the final contract to produce the radar was not even awarded until that year. The radar would have been better named the WSR-96, because the final installation was not completed until then. It almost didn't arrive in the twentieth century at all!

When the NEXRAD contract was finally given to Unisys, Burgess moved to the NWS's new NEXRAD Operations Group and was named Chief of the Operations Branch. His group oversaw the installation and training for the initial WSR-88D deployments and worked to improve the radar as much as science and computational power would allow.

Lemon worked with Unisys to apply and design the NWS's specifications into the new radar. Things were progressing until—for reasons unknown—the NWS decided to test Unisys to determine whether Unisys was out of compliance with the terms of its contract.

The NWS required Unisys to produce and install a working radar in either Connecticut or Oklahoma on April 20, 1991. Lemon picked Norman because severe weather was a sure thing in Oklahoma in spring, and he was confident the radar would perform.

Unisys scrambled to get the radar in place, and it was activated on April 20 at the NWS office at Will Rogers World Airport in Oklahoma City. Under the terms of the test, the NWS was not supposed to use the radar for real-world issues; it was there for test and demonstration purposes only, and its display was intentionally put in a remote part of the forecast office.

On April 21 central Oklahoma experienced the worst outbreak of severe weather in three years. The meteorologists at the Will Rogers office found the new radar so useful that they were literally running back and forth from the new Doppler radar to their conventional equipment. The radar had proved itself in its initial test.

Five days later, on April 26, 1991, the new radar in Oklahoma again performed flawlessly, providing data on three huge storm systems: the Wichita-Andover tornado (F5), the Cowley County tornado (F4), and the Red Rock Oklahoma tornado (F5).

In spite of this success, the NWS and Unisys were still not in sync. The NWS even made noises about restarting the procurement process, which would have delayed the installation of the radars by at least another three to five years. Remember, it had already been eighteen years since Doppler had proved itself in the Union City tornado. And by now the existing WSR-57s were falling apart.

Fortunately, two things kept the procurement process from starting up again. NWS employees using the radar in Oklahoma did their own evaluation of its performance and threatened to release this to the public. In addition, Unisys went on a political offensive on Capitol Hill. It was so effective that during a Congressional hearing regarding the major tornado outbreak of April 26, 1991, Kansas Congressman Dan Glickman exclaimed, "I want my NEXRAD!"

No one seems to have a definitive answer as to why the acquisition of the radar was so tortuous, even by government acquisition standards. So let's be grateful the radars were installed with no additional delays.

WeatherData was the first end-user organization in Kansas (and one of the first in the nation) to have access to the Doppler wind data. Our first day with this access we were watching a storm near the Santa Fe Railroad track south of Norman, Oklahoma. I was staring at the screen along with my colleague Ken Smith, and we both thought we saw a signature of a developing tornado. Because the storm was not in the conventional configuration, and because the early NEXRAD software was minimally helpful, we bent over to our left sides and turned our heads upside down so that the storm would be in the same configuration as the storms viewed in our training.

It appeared that these storms had the swirl of a developing tornado. So we issued our first-ever Doppler-based tornado warning, as the storm did not have a hook.

The Santa Fe Railroad stopped a train based on that warning. Minutes later, an error message flashed at the signal desk in the railroad's operations center in Schaumburg, Illinois. A tornado had crossed the track, taking out a trackside signal light and causing the error message. But because of the Doppler wind data and our resulting warning, there was no train in the area to be hit by the tornado. The value of Doppler radar had proved itself.

In some places, however, the Doppler radar was more controversial. Both the *Wall Street Journal* and *Time* published articles about storm warning accuracy declining after the first of the radars had been installed. By one measure of storm warning accuracy, the false-alarm rate, this was true. Because the radar could see small as well as large swirls of air, too many tornado and severe thunderstorm warnings were issued in the early days of the radar.

In retrospect, this result was not surprising. When more powerful medical imaging devices were first used by doctors in the 1990s, the number of "suspicious" images of potential breast cancer increased. So did the number of false diagnoses of prostate cancer when new tests emerged. With new technology, you don't know what you don't know. So the understandable err-on-the-side-of-safety result was to issue a tornado warning when new technology indicated a problem might exist. Once we gained experience with Doppler radar, the false-alarm rate substantially declined, but to this day, it is still somewhat higher than we would like.

Throughout the 1990s and 2000s, Doppler data, communications, and software have continued to improve. Doppler radar is still not perfect: there are times when it cannot detect a tornado or it seems to detect a tornado that does not exist. The false-alarm rate is still too high. When used to measure hurricane wind speeds, there are times the radar results are ambiguous. And the highest winds in tornadoes and hurricanes remain difficult to measure.

However, in the fifteen years since the first installations, Doppler radar has proved itself useful over and over again. Considering the countless lives it has saved and the millions of dollars in property loss it has prevented, a reasonable case can be made that the Doppler radar program is the single most successful scientific program, in terms of value (cost/benefit), ever undertaken by the federal government.

In 1995, Don Burgess became chief of the NEXRAD training branch. He has since retired from full-time employment at NSSL but still works there part-time. Severe storm science is in Burgess' blood, and we owe him a debt of gratitude for his four decades of work to give meteorologists the means to consistently make effective storm warnings.

Les Lemon has gone back to work in Norman for the National Severe Storms Lab.

Anyone can go to accuweather.com and pull up the data from any radar in the United States for zero cost (the site is supported by advertising). Or you can go to the NWS itself for radar data. WeatherData and the NWS both have products that meld data from radar and 11,000 rain gauges to create a highly accurate rainfall map every 24 hours. Airlines and the FAA use radar to route planes around storms. Hurricane warnings are more accurate because radar allows a better assessment of their risks. The benefits of the NEXRAD radar are countless.

We have come a long way since the early 1990s. While it was a long and harder-than-it-needed-to-be journey, the NWS finally got NEXRAD right and we are all better off for it. Clearly, for its cost, Doppler radar is one of the best investments the federal government has ever made.

CHAPTER EIGHTEEN

HURRICANE ANDREW

WEATHERDATA'S RAILROAD BUSINESS, WHICH BEGAN
with our forecasting for Southern Pacific's Cotton Belt line from
Kansas City to Tucumcari, New Mexico, in 1986, grew rapidly. After
our first-day accurate warning prevented a tornado from hitting a
Southern Pacific train, word of our effective forecasting spread, and
individual Southern Pacific railroad divisions signed up for our service. By 1988 we were serving the entire Southern Pacific system,
including its Sunset Route that ran from Los Angeles to New Orleans,
with stops at Houston and Beaumont, Texas. We also provided data
for the Southern Pacific lines that ran southwest from Houston along
the Texas coast to Brownsville. These lines had special needs: they
needed hurricane warnings.

But there was one problem: We had never issued formal hurricane warnings. So I went back to the textbooks. And luckily, a couple
of WeatherData team members had hurricane experience, so we put
together a program in which we had a reasonable degree of confidence.

I didn't know our new program was going to get a major test that first season, though.

Hurricane Gilbert, in September 1988, strengthened faster than any other hurricane in history as it moved through the Gulf of Mexico on a course for the United States. The barometric pressure dropped to 26.22 inches, the lowest atmospheric pressure reading in the history of the western hemisphere. A hurricane hunter aircraft measured a wind gust of 199 miles per hour, as strong as the winds of many of the strongest tornadoes. In addition to Gilbert's highly damaging winds, its storm surge, if the hurricane stayed at Category 5, would be devastating.

As you can imagine, Southern Pacific's interest in the forecast path of Gilbert was great. They could lose millions of dollars if we got the hurricane's projected path wrong.

We examined all of the pertinent data carefully. The three-day Spectral model, a computerized simulation of the atmosphere that, at the time, was used by meteorologists to forecast hurricane paths, forecast a weakening in a high-pressure system that might turn Gilbert toward the upper Texas coast. Given the high populations of Houston and Beaumont, this super hurricane would cause incredible devastation if it struck this region at its peak intensity.

We heard via radio that one of our competitors was banking on the Spectral model and had forecast Gilbert to strike the upper Texas coast. That started the WeatherData phones ringing. "Why are you forecasting Gilbert to hit the lower Texas coast when the radio says the hurricane is going to hit Houston?"

We at WeatherData believed Gilbert would not turn north and instead stay more or less on its established path, striking the lower Texas coast around South Padre Island where the population is less dense. We also thought Gilbert would weaken. That is the forecast we gave Southern Pacific, and we stuck with it. After conferring with us several times, Southern Pacific decided to continue operations from Houston to New Orleans and from Houston west to California. The other railroads—those that didn't subscribe to WeatherData—shut down; one even boarded up its regional office in Houston.

It turned out that our first-ever critical hurricane forecast was a success. A weakened Gilbert struck northern Mexico just south of the mouth of the Rio Grande. Some strong winds occurred on Padre Island and around Brownsville. Houston and New Orleans were hit with some breezy rain showers. That's it.

After this success, our railroad business expanded with the addition of the Santa Fe Railroad in 1989, and both the Kansas City Southern and the Denver and Rio Grande Western in 1991. Our newspaper business was growing as well. Railroads and newspapers in hurricane-prone areas wanted detailed hurricane forecasts; we continued to improve our skills in tropical meteorology and, equally important, the technology to convey critical hurricane information.

With the addition of the *Miami Herald* and the *New Orleans Times-Picayune* as clients, the frequency of client-requested hurricane forecasts increased. Our work for the *Herald* went especially well, even bringing a complimentary letter from the chief meteorologist for the NWS in Miami.

* * *

Newspaper weather was focused on the overall forecast rather than storm warnings. When we signed the contract with the *Miami Herald*, it never occurred to us that we might have to advise the Miami area about a storm with the intensity of Hurricane Andrew (1992).

To understand the events around Hurricane Andrew and why it was pivotal in the fields of insurance and emergency management and helped develop an entirely new professional field called "business continuity," it's necessary to talk about hurricane science and how the United States viewed hurricanes in the early 1990s.

While tornadoes have more *concentrated* energy, it is hurricanes that have the potential to cause the worst destruction and greatest loss of life in the United States. It is estimated the Galveston hurricane of 1900 caused from 8,000 to 12,000 deaths, and the 1928 Okeechobee hurricane caused at least 2,500 and perhaps as many as 6,000 deaths. But as hurricane hunters, satellites, and other technology were brought to bear on the hurricane forecast and warning problem, the death rate from hurricanes began to decrease. The period from 1970 to 1988 saw fewer hurricanes than average strike the United States, and all of the hurricanes that came ashore were Category 3 or lower.

Why is that important? Wind *force*, the destructive power of the wind, is a square function rather than a linear one. So a 150-mile-per-hour wind isn't twice as powerful as a 75-mile-per-hour wind; it is four times more powerful. If every other factor is equal, for every one dollar of damage caused by a Category 1 hurricane, a Category 2 will cause 10 dollars; a Category 3 will cause 50 dollars; a Category 4 will cause 250 dollars; and a Category 5 will cause 500 dollars in damage.

Given the availability of effective air conditioning (rare and expensive prior to the 1960s) and America's growing prosperity, what *USA Today* called a "lifestyle migration" began. People began flocking to the beaches and cities along the Gulf and Atlantic coasts in the 1970s and 1980s. Businesses moved south. Miami became the capital of inter-American trade and culture.

A number of coastal counties saw their populations increase fivefold, sixfold, or sevenfold or more in just twenty years. And during that time, the threat of hurricanes was largely forgotten, because the number of hurricanes during the 1970s and 1980s was lower than average in the United States.

But a number of meteorologists were becoming increasingly alarmed about the complacency mind-set regarding hurricanes. Bob Sheets, the director of the National Hurricane Center, gave numerous cautionary presentations about the possible devastation and dangers of Category 4 and Category 5 hurricanes. So did Bill Gray, a hurricane expert from Colorado State University.

One can think of the hurricane warning and mitigation process as taking place in three steps:

1. The forecast,

2. Action taken to protect life (evacuation) and property (boarding up), and

3. The post-storm aftermath and recovery.

Meteorologists are the primary drivers of the forecast and have influence on how smoothly the run-up to the hurricane plays out. But meteorologists have almost no influence over the recovery phase of a hurricane.

Meteorologists saw Andrew begin as a low-pressure wave (a broad area of relatively low atmospheric pressure lacking a well-defined center around which the wind spins in a complete circle) that moved west into the Atlantic from the coast of Africa. Once the wave moved into the Atlantic, the tracking and forecasting process began.

The first step in making a forecast is to locate any hurricanes or tropical storms already in existence. The second step is to forecast their future path. The third is to forecast their changes in intensity. Once the changes in intensity (which meteorologists define in terms of atmospheric pressure and wind speed) are complete, other vital forecasts such as rainfall amounts and storm surge can be made. Then, the final forecast is communicated to the NWS field offices and local emergency management officials so evacuations and other preparatory actions can begin.

On August 14, 1992, a low-pressure system moved west from Africa and crossed into the Atlantic. At first there was nothing that would differentiate this tropical wave from the dozens of others that move from the African coast each summer.

A week later, that changed.

High pressure formed over the southeast United States, and the wind pattern six miles up in the atmosphere shifted to allow the rising air in the center of the storm to better ventilate. Like a supercell thunderstorm, a hurricane must be ventilated in order to grow and survive. Andrew reached hurricane strength the morning of August 22 and started to intensify toward Florida and the Bahamas, heading west at 18 miles per hour.

WeatherData was keeping in touch with the *Miami Herald* during this period as Andrew rapidly strengthened and continued to move toward south Florida. The hurricane had reached Category 4 intensity when it passed over Eleuthera Island in the Bahamas. At this point, it looked like a major hurricane was inevitable in south Florida. We were convinced the storm would be extremely severe and that it would strike Miami, so I came in on a Sunday, the day before we were predicting landfall to occur, to help out our meteorologists.

One of our contacts at the *Herald* called about four o'clock in the afternoon and asked if there was any chance Andrew would miss. I told him no and added that I thought it would be "the most costly disaster ever to hit the United States." He seemed very surprised by my words.

I'd gotten involved in the hurricane disconnect problem (the debate between meteorologists and those outside the meteorology profession about the potential threats of hurricanes) shortly after Gilbert, when some of our insurance clients asked me to do consulting work regarding worst-case tornado and hurricane scenarios. One thing led to another and I was asked to be a speaker at the National Association of Mutual Insurance Companies' annual meeting in Orlando in 1991. At that time, there was a great deal of controversy within the insurance industry as to how high the losses could be if a major hurricane hit the United States.

There were many people within meteorology who thought that, given the explosion of affluent population along the Gulf and Atlantic coasts, there was some likelihood that a major hurricane could cause several tens of billions of dollars in damage; I said so at the convention. I also showed a map of the paths of hurricanes from the 1950s and 1960s that featured many Florida landfalls, stressing that similar hurricanes could strike again (and they did in 2004 and 2005), and that since Florida's coastal population had increased by a factor of ten since the 1950s, large damage tolls would result. Finally, I discussed Dr. William Gray's research regarding a 40-year hurricane cycle, and that we were currently in the low part of the cycle. I said that when the higher part of the cycle returned, as it likely would in the '90s and 2000s, we would inevitably see more damage than the insurance industry was used to experiencing.

I had spent a great deal of time and effort to research and prepare my speech, and I hoped for a favorable reaction. The people who invited me to speak were complimentary about the quality of the speech and the way I delivered it, but the general reaction was less enthusiastic than I'd expected.

It turns out that many people in the audience just didn't believe me. Camille, a Category 5 hurricane, had hit a populated area (the Mississippi Gulf Coast) in 1969 and caused $1.42 billion in damage, so how could a hurricane cause tens of billions of dollars in damage?

I was remembering that meeting of insurance companies and the great skepticism many people held regarding the destructive potential of hurricanes as I was watching the evening news coverage that Sunday as Andrew approached the Florida coast. The warnings from the National Hurricane Center (NHC) were, I believed, excellent. They aligned with the forecasts we had given our clients. But if the forecasts turned out to be correct, I was curious as to what the storm's ultimate toll would be. I set my alarm clock for 4:00 a.m., as I wanted to be in the office when Hurricane Andrew made landfall. I was expecting that we would be overwhelmed with calls.

The NWS Hurricane Andrew warnings for the public were excellent. Here is just one example:

BULLETIN ... HURRICANE ANDREW INTERME-
DIATE ADVISORY NUMBER 30A ...

NATIONAL WEATHER SERVICE MIAMI FL . . .

8 PM EDT SUN AUG 23 1992 . . .

EXTREMELY DANGEROUS HURRICANE ANDREW
BEARING DOWN ON SOUTHEAST FLORIDA . . .
HURRICANE WARNINGS REMAIN IN EFFECT
FOR THE CENTRAL AND NORTHWEST BAHA-
MAS . . . THE FLORIDA EAST COAST FROM VERO
BEACH SOUTHWARD THROUGH THE FLORIDA
KEYS TO THE DRY TORTUGAS . . . THE FLORIDA
WEST COAST SOUTH OF VENICE . . . AND FOR
LAKE OKEECHOBEE. A HURRICANE WATCH
AND A TROPICAL STORM WARNING ARE IN
EFFECT FOR THE FLORIDA EAST COAST FROM
VERO BEACH NORTHWARD TO TITUSVILLE . . .
AND ON THE FLORIDA WEST COAST NORTH OF
VENICE TO BAYPORT. ALL PRECAUTIONS TO
PROTECT LIFE AND PROPERTY . . . INCLUDING
EVACUATIONS ORDERED BY EMERGENCY MAN-
AGEMENT OFFICIALS . . . SHOULD BE RUSHED
TO COMPLETION.

AT 8 PM EDT . . . 0000Z . . . THE CENTER OF
ANDREW WAS LOCATED NEAR LATITUDE 25.4
NORTH . . . LONGITUDE 77.3 WEST OR ABOUT
185 MILES . . . 300 KM . . . EAST OF MIAMI
FLORIDA. ANDREW IS MOVING TOWARD THE
WEST NEAR 16 MPH . . . 26 KM/HR . . . AND
THIS MOTION IS EXPECTED TO CONTINUE FOR
THE NEXT 24 HOURS. HURRICANE CONDITIONS
ARE SPREADING OVER THE NORTHWEST AND
CENTRAL BAHAMAS AT THE PRESENT TIME.

A WIND GUST TO 120 MPH WAS REPORTED AS
THE CENTER PASSED OVER THE NORTHERN
END OF ELEUTHERA ISLAND. ON THE PRESENT
COURSE . . . TROPICAL STORM FORCE WINDS
SHOULD BEGIN ON THE SOUTHEAST FLORIDA
COAST AFTER MIDNIGHT . . . WITH HURRICANE
CONDITIONS EXPECTED IN THE PREDAWN
HOURS.

THIS IS A DANGEROUS CATEGORY FOUR HUR-
RICANE WITH MAXIMUM SUSTAINED WINDS
NEAR 145 MPH . . . 235 KM/HR . . . IN A SMALL
AREA NEAR THE CENTER. SOME FLUCTUATIONS
IN STRENGTH ARE LIKELY BEFORE LANDFALL.
HURRICANE FORCE WINDS EXTEND OUT-
WARD UP TO 30 MILES . . . 45 KM . . . FROM
THE CENTER . . . AND TROPICAL STORM FORCE
WINDS EXTEND OUTWARD UP TO 105 MILES . . .
165 KM.

Andrew was an extremely strong storm that eventually reached Category 5 intensity; but it was also very small. Note, above, that the hurricane-force winds extended only thirty miles from the hurricane's center. This would be crucial in the hours and days ahead.

As Andrew blew though southern Dade County, the National Hurricane Center (then located in the southwest part of Miami in a high-rise office building) was in hunker-down mode. Special shutters covered the windows. The families of forecasters were spending the hurricane in the administrative offices so the forecasters could do their work without distraction. Emergency generators were providing power in case the commercial power failed. A small team of forecasters had been evacuated to Washington, D.C., in case the hurricane center was knocked out of commission. The forecasters were as prepared as they could be.

The winds rose throughout the night. Just after 4:00 a.m., a gust of wind 107 miles per hour was recorded at the NHC as the eye crossed the coast into the south Miami area. A satellite antenna blew off the building a few minutes later.

At 4:52 a.m., the wind instrument recorded a wind gust of 164 miles per hour when the instrument failed as a distinct *thud* shook the building. The huge WSR-57 radar had been wrenched off the roof. Had the wind instrument survived, it is likely it would have recorded even higher gusts. Based on the final radar image of the hurricane, the meteorologists at the NHC knew the wind speeds had been even higher just south of their Dixie Highway location.

At the same time that the NHC was experiencing its highest winds, my 4:00 a.m. alarm went off. I went in to the office and there were surprisingly few client calls for what I believed would ultimately be a tremendous disaster. The pace at the office was so slow that I decided to go back home around 8:00 a.m., relax a little, then put on my suit and go back in for the day.

When I got home, I learned the reason the phone wasn't ringing at WeatherData. I tuned in to the local CBS station because I wanted to see Dan Rather's hurricane coverage. Rather had come to national attention for his coverage of 1961's Hurricane Carla while working for a Houston TV station. I thought he would probably provide the most insightful coverage of any of the networks.

To my surprise, Rather was standing outdoors north of downtown Miami, describing the partly sunny skies and gusty winds (not uncommon after a hurricane) and talking about how Miami had dodged a bullet.

I was shocked. Hurricane Andrew's eye had passed well south of downtown Miami, so that was where the worst damage would be, not north of downtown. It was the same on the other networks. The national coverage was describing far, far less damage than I had predicted, and I was starting to have a few doubts about my forecast. But I still thought that our WeatherData forecast had been correct, as Andrew's wind field (the distribution of winds within a storm), while unbelievably intense, was very small and certainly didn't extend into north Miami. The networks' evening newscasts, to an extent, repeated the "dodged a bullet" theme.

Tuesday morning, much of the national media boarded flights at Miami International Airport back to New York, Atlanta, and the other cities where they were based, thinking they had done their job, not realizing they had completely missed the disaster story of the decade.

* * *

In south Dade County, largely unknown to the outside world, the aftermath of Hurricane Andrew looked like a nuclear bomb had exploded: homes flattened, roads blocked, no power, no air conditioning, no running water.

And no one outside of the immediate area knew. There was no telephone service, either landline or cellular.

Because no one knew about the devastation, help was very slow in arriving. But the help was sorely needed: 28,000 buildings had been destroyed and another 110,000 had been damaged.

Back in Wichita, I was increasingly frustrated by the national news coverage. The destruction wrought by Hurricane Andrew was being played down; when pictures were shown, the damage was far less than I would have expected of a hurricane of that intensity. It just wasn't adding up.

In south Dade County, the nightmare worsened. Showers and thunderstorms moved through the area, causing further damage to homes with open roofs. There was no food, no water, no medical attention. The State of Florida's response was completely inadequate, mainly comprised of sending a few National Guard troops to the area.

State officials then told the Feds they didn't need federal troops. When the on-scene locals (Dade County officials) told the Feds otherwise, the Feds began playing bureaucrat. In those days, the Federal Emergency Management Agency (FEMA) was well known as a dumping ground for political cronies and was widely considered to be a joke. Officials in south Dade were getting the runaround from FEMA, in part because FEMA was getting differing information and requests from the Florida government.

Meanwhile, the residents of the 138,000 damaged buildings continued to swelter and suffer. Local Miamians, as well as citizens from Fort Lauderdale and West Palm Beach, attempted to drive bottled water, food, clothes, and other supplies into the stricken area. They were successful during the first few hours; but they were shut down by the end of the first day. The same bureaucrats who were giving mixed signals to Washington seemed to prefer organized suffering to disorganized relief.

I went home from work three and a half days after Andrew's landfall and turned on the news. I saw an obviously haggard, exhausted, and annoyed emergency manager named Kate Hale on the news, pleading for federal help. "Where's the cavalry?" she pleaded. Because her words, her obvious exhaustion, and her frustration were so authentic, everyone watching knew the situation was dire. Aerial photographs from news helicopters revealed, finally, mile after mile of flattened homes, power poles, and mobile homes. President George H. W. Bush ordered FEMA into action, and relief started to arrive, in quantity, on Saturday.

This was five days after the hurricane had left the area.

The injuries caused by the hurricane continued to occur months after the storm itself was a memory. Unscrupulous contractors preyed upon storm victims. Local government's reaction was to crack down on out-of-town contractors through a system of licensing contractors. This new process could have worked just fine, but months passed before the first licensing test was even administered. Given the overwhelming demand for builders and contractors, this slowed the recovery even further. Hundreds of businesses, maybe more, closed, never to reopen.

But Hurricane Andrew was a triumph for meteorology: the forecasts of this extreme hurricane were extremely accurate. The staff of the NHC performed heroically; they not only made accurate forecasts as Andrew approached Florida, they continued to track Category 3 Andrew across the Gulf of Mexico and into a sparsely populated part of Louisiana where the hurricane made its final U.S. landfall. The death toll from Andrew was an unbelievably low 24, and that number would have been smaller still had timely medical aid been available for those in need in south Dade County.

But the aftermath of the disaster was, well, a complete disaster. Why?

For nineteen years, not a single hurricane greater than Category 3 intensity had made landfall in the United States. When Category 4 Hugo caused that streak of good fortune to end in 1989, the damage it caused was relatively minor (for a Category 4 hurricane) because its highest winds had struck the least populated point on the East Coast. So in spite of excellent forecasts and storm tracking by the hurricane center and by the private meteorological community, few national first responders (FEMA, insurance companies, network news departments, etc.) had any idea of what Andrew, a Category 5 storm was capable of. Almost no one had had firsthand experience twenty-three years earlier with Hurricane Camille, the last comparable hurricane to strike the United States. *And* many people believed they understood Category 4 storms, based on their experience responding to Hugo in nearby Charleston, which didn't experience anything near the storm's highest winds.

Even though a number of us were trying to raise awareness, one might say there was a "failure of imagination" regarding what a strong hurricane could do.

One group that *did* learn the lessons of Andrew was the business community.

As computerization became more and more important to American business, some forward-thinking companies realized that a loss of power could be catastrophic. In the old days, if the power went out, manual typewriters and adding machines still worked. If the air conditioning failed, people were uncomfortable, but work could proceed.

But once a business was computerized, a power failure meant that:

- Word processors didn't work.

- Computers didn't work.

- Lack of air conditioning could jeopardize delicate computer components.

- Crashing hard discs could mean the loss of vital corporate records.

Thus it was that the field of business continuity was born in the 1980s. *Disaster Recovery Journal*, a trade publication for the business-continuity community, had its first conference and trade show in Atlanta in 1989. Business-continuity departments began the process of organizing corporate computer backups, alternate sites, human resource planning (if your employees' homes are damaged they cannot come to work), and other measures. Once the degree and extent of the devastation of Andrew became well known, business-continuity departments moved up in influence and prestige in many corporations.

Another industry that made major changes was the insurance industry. Eleven insurance companies went bankrupt after Hurricane Andrew, and thirty-three companies were allowed by state regulators to stop writing insurance in Florida. In the wake of Andrew, insurance companies began using commercial meteorology companies, risk management models, and new financial instruments such as "hurricane bonds" (bonds that paid a high rate of interest in return for the use of the principal amount to pay premiums in the event of a catastrophic hurricane). Hurricane Andrew prompted a major change in the financial foundations of the property and casualty insurance industry.

Meteorologist Bob Sheets wrote in *Hurricane Watch* that he was worried about Andrew striking New Orleans as it made its way across the Gulf of Mexico after passing through Florida. Meteorologists knew very well that if a major hurricane were to strike below-sea-level New Orleans, the death toll could be in the thousands or even tens of thousands.

Andrew moved ashore in a swampy part of Louisiana. Its second U.S. landfall was anticlimactic. It would take another thirteen years before Bob Sheets' nightmare would be realized in the form of Hurricane Katrina.

And, like Andrew, meteorology would do a great job with the forecasts of Katrina. And also like Andrew, Katrina's aftermath would be a far worse disaster than the storm itself.

CHAPTER NINETEEN

KATRINA: PART ONE

CUMULUS CLOUDS CLUSTERED OVER THE CLUB
Peace and Plenty Resort on Exuma Island, August 23, 2005. At first,
the clouds appeared identical to the clouds that fill the sky 339 days a
year in the languid out-islands of the Bahamas.

But these clouds were different. Earlier in the day they'd begun a
subtle swirl, invisible to the guests visiting Club Peace. To the vaca-
tioners, the clouds—and the showers they brought—were merely
obstacles to a suntan. Over a lunch of the house specialty, conch burg-
ers, guests chatted about when the weather would clear and when the
steady breeze would calm.

It was an unlikely beginning to what would become in both human
suffering and economic terms the worst natural disaster in contem-
porary North America. If Andrew represented the very best in what
hurricane warnings could do in terms of saving lives, why did so many
die in Katrina?

The answers are complex but important, and there's plenty of blame to go around. Katrina was hardly the worst-case meteorological scenario, as I will explain in the following pages. If similar mistakes are made in future major storms, the resulting human suffering will be the same as it was in Louisiana and Mississippi in 2005.

AUGUST 23, 2005

By 2005, the National Hurricane Center (NHC) had moved from the high-rise on Dixie Highway to a low-rise steel-and-concrete-reinforced bunker-like structure at Florida International University. It was located on Miami's west side, far away from the coast.

The design of the building was heavily influenced by the center's experience during Hurricane Andrew. Since wind speeds increase rapidly with height above the ground (there's less friction to retard wind speeds), the center is a sprawling single-story structure. While the location and architecture had changed since Andrew, the mission of the NHC and the basic techniques of tropical weather forecasting had not.

After a review of the day's data, the NHC designated the developing weather system in the Bahamas to be the year's Tropical Depression Twelve.

The 2005 hurricane season, which began, as is traditional, on the first of June, had been much busier than usual through the first half of August. When hurricanes threaten the U.S. mainland, there's a plethora of media crowding the desk where the director of the hurricane center sits during press briefings. Yet there was little media activity at the NHC the afternoon of August 23.

The national media was worn out from covering the highest number ever of tropical storms for so early in a hurricane season and was looking for fresh topics. So the *New York Times* had nothing about the developing storm on its August 24 front page. The *Wall Street*

Journal likewise did not mention Tropical Depression Twelve on its front page.

As the incipient Katrina developed, what *were* the newspapers covering? Cindy Sheehan, a woman who had lost her son in Iraq, was camped out in Crawford, Texas, wanting to have a second meeting with President Bush. Other newspapers were covering, in sometimes breathless terms, MTV's Video Music Awards that were going to be presented Sunday evening in Miami.

AUGUST 24, 11:00 A.M.

```
TROPICAL STORM KATRINA DISCUSSION NUMBER
4 NWS TPC/NATIONAL HURRICANE CENTER
MIAMI FL 11 AM EDT WED AUG 24 2005 . . .
SATELLITE IMAGERY . . . DOPPLER RADAR
DATA FROM THE BAHAMAS AND MIAMI . . .
AND RECONNAISSANCE WIND DATA INDICATE
TD-12 HAS BECOME MUCH BETTER ORGANIZED
THIS MORNING AND HAS STRENGTHENED INTO
TROPICAL STORM KATRINA.
```

In Miami, the hurricane center was gearing up for a higher level of activity and a potential onslaught of media, since it looked like Katrina was going to continue to intensify and might move into densely populated south Florida.

When a major hurricane threatens the United States, the "lead" hurricane forecaster position becomes a pressure cooker of a job with public, media, and political interest intensely high, and that interest often extends all the way up to the president of the United States.

At the time, the NHC looked on the media as a mixed blessing. The media was essential to getting the word out to the public, but a number of the regular reporters assigned to the pool coverage tended to be disruptive. The media area at the NHC is not especially large, so space is at a premium. Pool coverage is a system where the networks share certain technical equipment (cameras, cables, microphones, etc.) to make coverage practical for all of the stations wanting access. The 2004 hurricane season had been the most destructive since 1992's Andrew, especially in Florida as multiple storms struck the state. So in 2005, there was intense public interest whenever a storm might hit the United States.

With the advent of the Weather Channel and the realization that its ratings and the number of people viewing its website soared during hurricanes, every cluster of clouds that moved west from the African coast into the Atlantic in summer was the subject of intense speculation as to whether it might turn into "something." In the twenty-first century, it seems that for every actual hurricane, we get three days of genuine watches and warnings preceded by ten days of anxiety.

Katrina, however, was different. She developed fairly close to the U.S. coast. So the hurricane center was already facing the likelihood of issuing watches and warnings for Florida almost from the time Katrina became a tropical storm.

AUGUST 24, 1:00 P.M.

The South Beach section of Miami Beach was gearing up for the live telecast of the 2005 MTV Video Music Awards. Whether it's the Super Bowl, the Grammys, or the Video Music Awards, this type of glitzy, high-profile event is the lifeblood of the South Beach tourism industry, and the area's already pricey hotel rates had soared to more than $10,000 per night in some venues. As a symbol of MTV's conquering of South Beach, a giant inflatable astronaut, the "Moonman" MTV logo, was hoisted atop the Surfrider Hotel.

So, the Video Music Awards staff at South Beach was closely monitoring the NHC forecasts concerning tropical storm Katrina—the
live telecast awards ceremony was just four days away. The media pre-
ferred to spend its time at glitzy South Beach rather than in the press
pool at the NHC.

Over the warm Atlantic waters, Katrina continued to get better
organized, and her wind speeds increased accordingly.

Pre-award outdoor parties for the Video Music Awards were
clearly threatened by the newly named Katrina. With each passing
hour, the forecast weather looked worse and worse. The South Beach
Chamber of Commerce was especially concerned about the effects of
Katrina, since the previous year's award festivities had been dampened
by tropical winds and rains. A second foul-weather disruption could
threaten Miami's viability for future award ceremonies.

BULLETIN: TROPICAL STORM KATRINA INTER-
MEDIATE ADVISORY NUMBER 4A

NWS TPC/NATIONAL HURRICANE CENTER MIAMI
FL . . . 2 PM EDT WED AUG 24 2005 . . .

TROPICAL STORM KATRINA STRENGTHENING
OVER THE CENTRAL BAHAMAS . . . HEAVY RAIN-
FALL THREAT FOR THE BAHAMAS TONIGHT AND
THURSDAY . . . A TROPICAL STORM WARNING AND
A HURRICANE WATCH REMAIN IN EFFECT FOR
THE SOUTHEAST FLORIDA COAST FROM VERO
BEACH SOUTHWARD TO FLORIDA CITY . . . THIS
REPLACES THE TROPICAL STORM WATCH . . . A
HURRICANE WATCH MEANS THAT HURRICANE
CONDITIONS ARE POSSIBLE WITHIN THE WATCH
AREA GENERALLY WITHIN 36 HOURS.

As Katrina strengthened, the inconceivable occured. The giant inflatable Moonman on top of the Surfcomber Hotel would have to come down. The area was now under a hurricane watch.

AUGUST 25, 8:00 A.M.

BULLETIN: TROPICAL STORM KATRINA INTERMEDIATE ADVISORY NUMBER 7A

NWS TPC/NATIONAL HURRICANE CENTER MIAMI FL . . . 8 AM EDT THU AUG 25 2005

KATRINA SLOWLY GETTING BETTER ORGANIZED AS IT MOVES WESTWARD . . . TOWARD SOUTHEASTERN FLORIDA . . . A HURRICANE WARNING REMAINS IN EFFECT FOR THE SOUTHEAST FLORIDA COAST FROM VERO BEACH SOUTHWARD TO FLORIDA CITY INCLUDING LAKE OKEECHOBEE. A HURRICANE WARNING MEANS THAT HURRICANE CONDITIONS ARE EXPECTED WITHIN THE WARNING AREA WITHIN THE NEXT 24 HOURS. PREPARATIONS TO PROTECT LIFE AND PROPERTY SHOULD BE RUSHED TO COMPLETION.

While the MTV people were scrambling at South Beach to cope with the deteriorating weather and weather forecasts, the National Weather Service was mobilizing. Forecasters knew that Katrina might well be a minimal hurricane when she struck the Florida coast; they were far more worried about what might happen when she moved into the Gulf.

Water temperatures, correlated to hurricane strength (warmer water temperatures make hurricanes stronger), were at near-record levels. Wind shear five miles above the ocean was very low, and lighter wind shear means the potential for stronger hurricane winds. It looked like a perfect situation for major strengthening. So the hurricane center sent for reinforcements.

MTV's Moonman is lowered to the ground in advance of Katrina. Photo by Daniel Hubalek.

It is impractical and inefficient to constantly staff the National Hurricane Center with all the people needed to run the center when a *major* hurricane threatens the United States. The media—Spanish as well as English—emergency managers, politicians (local, state, and federal), FEMA, and countless others demand to be kept informed and in the loop. When the NHC believes a major storm will develop, calls are placed to trained, experienced meteorologists from NWS offices across the nation to get on planes bound for Miami.

By law the Federal Emergency Management Agency (FEMA) is supposed to play the key role in mitigating major disasters. So the NWS established the FEMA Hurricane Liaison Team (HLT) to help ensure a quick and appropriate response when a major storm threatens. One of the individuals on the liaison team who received the call for Katrina was Bill Read, the meteorologist-in-charge of the NWS in Houston. A savvy meteorologist who had handled numerous major storms, Read started making quick plans to get to Miami.

AUGUST 25, 3:00 P.M.

BULLETIN . . . TROPICAL STORM KATRINA INTERMEDIATE ADVISORY NUMBER 8B

NWS TPC/NATIONAL HURRICANE CENTER MIAMI FL . . . 3 PM EDT THU AUG 25 2005 . . .

KATRINA JUST BELOW HURRICANE STRENGTH AS IT MOVES SLOWLY WESTWARD ACROSS THE FLORIDA STRAITS TOWARD SOUTHEAST FLORIDA . . . TROPICAL STORM FORCE WINDS NEARING THE FLORIDA COAST . . . A HURRICANE WARNING REMAINS IN EFFECT FOR THE SOUTHEAST FLORIDA COAST FROM VERO BEACH SOUTHWARD TO FLORIDA CITY . . . INCLUDING LAKE OKEECHOBEE.

A HURRICANE WARNING MEANS THAT HURRICANE CONDITIONS ARE EXPECTED WITHIN THE WARNING AREA WITHIN THE NEXT 24 HOURS. PREPARATIONS TO PROTECT LIFE AND PROPERTY SHOULD BE RUSHED TO COMPLETION. 3

PM EDT REPORTS FROM NOAA DOPPLER RADARS
AND A NOAA RECONNAISSANCE INDICATE THE
CENTER OF TROPICAL STORM KATRINA WAS
ABOUT 35 MILES EAST-NORTHEAST OF FORT
LAUDERDALE, FLORIDA.

Bill Read was starting to board his 3:00 p.m. flight from Houston
to Miami when he was held up by the gate agent. The flight to Miami
had just been cancelled due to the impending storm. Read scrambled
and found a flight to Orlando. Being an experienced meteorologist,
he knew that Katrina—at that time—was small enough for him to fly
safely into Orlando International farther north. His plan was to drive
to Miami that night.

Back at South Beach, parties were being called off or moved
indoors, and hotel reservations were being cancelled for the night.
Katrina was now a hurricane.

BULLETIN . . . HURRICANE KATRINA ADVISORY
NUMBER 9

NWS TPC/NATIONAL HURRICANE CENTER MIAMI
FL . . .

5 PM EDT THU AUG 25 2005 . . .

STRENGTHENING HURRICANE KATRINA BEAR-
ING DOWN ON THE SOUTHEAST COAST OF
FLORIDA . . .

It wasn't just Read who had travel problems. The weather dete-
riorated rapidly, much more rapidly than many local officials had
anticipated. The airport started to have problems as the winds rose
and the rains intensified. According to news reports, Lindsay Lohan

and Alicia Keys arrived at the Miami airport but had trouble getting out as Katrina moved in.

Power started to fail as the northern suburbs of Miami were buffeted by the rising winds. Rush hour in Miami-Dade County quickly became a quagmire as roads flooded and trees blocked highways. By 8:00 in the evening, Katrina was producing 80-mile-per-hour winds in north Miami with 92-mile-per-hour gusts at Port Everglades, just southeast of Fort Lauderdale.

Meanwhile, Bill Read had arrived at Orlando Airport, rented a car, and headed south on the Florida Turnpike. He planned to get all the way to Miami, but the severe weather conditions forced him to divert and stop at Jupiter for the night.

> HURRICANE KATRINA INTERMEDIATE ADVISORY NUMBER 9B
>
> NWS TPC/NATIONAL HURRICANE CENTER MIAMI FL
>
> 9 PM EDT THU AUG 25 2005
>
> ...KATRINA RELENTLESSLY POUNDING SOUTH FLORIDA ... CALM OF THE LARGE EYE EXPERIENCED AT THE NATIONAL HURRICANE CENTER ...

AUGUST 26

As far as Florida was concerned, Katrina had come and gone.

> BULLETIN ... HURRICANE KATRINA INTERMEDIATE ADVISORY NUMBER 11A

NWS TPC/NATIONAL HURRICANE CENTER MIAMI
FL . . . 7 AM EDT FRI AUG 26 2005 . . .

KATRINA CHURNING WESTWARD OVER THE
GULF OF MEXICO . . . AT 7 AM EDT . . . 1100Z . . .
THE CENTER OF HURRICANE KATRINA WAS
LOCATED NEAR LATITUDE 25.3 NORTH . . .
LONGITUDE 81.8 WEST OR ABOUT 50 MILES
NORTH OF KEY WEST FLORIDA AND ABOUT 60
MILES SOUTH OF NAPLES FLORIDA. KATRINA
IS MOVING TOWARD THE WEST NEAR 5 MPH
AND THIS MOTION IS EXPECTED TO CONTINUE
FOR THE NEXT 24 HOURS . . . WITH A SLIGHT
INCREASE IN FORWARD SPEED.

While most south Floridians were cleaning up the mess made by Katrina and trying to resume a normal routine, the MTV awards were quickly moving to regain lost ground.

* * *

By the time the 2005 MTV Video Music Awards telecast went on the air Sunday night, most of the population of New Orleans, 75 percent by some estimates, had left the city.

They joined residents of southern Alabama, southeast Louisiana, and Mississippi in one of the largest mass evacuations in history—one million in all. Many of them passed the time that Sunday evening watching weather reports, *Desperate Housewives*, and even the Video Music Awards on hotel and motel room televisions.

Other than meteorologists and disaster experts, little did anyone realize that, because of Katrina, many of the storm refugees would not be back anytime soon.

CHAPTER TWENTY

KATRINA: PART TWO— INACTION IN ACTION

FRIDAY, AUGUST 26

BILL READ AWOKE FRIDAY MORNING AND HEADED south from his hotel in Jupiter. He had to thread his way around downed trees and power lines in order to get to the National Hurricane Center. He knew he'd be needed in the hours and days to come. Upon his arrival, he found there was already a sense among the NHC staff that Katrina might be a historic storm. The population of the NHC forecasting and dissemination operation had grown from the usual two staffers to eleven.

BULLETIN . . . HURRICANE KATRINA ADVISORY NUMBER 14 NWS TPC/NATIONAL HURRICANE CENTER MIAMI FL . . . 5 PM EDT FRI AUG 26 2005 . . .

KATRINA CONTINUING TO MOVE WEST-SOUTH-
WESTWARD AWAY FROM THE FLORIDA KEYS . . .
WATCHES AND WARNINGS DISCONTINUED FOR
MAINLAND FLORIDA . . .

Katrina had moved off the west coast of Florida and was out in the Gulf. At face value, advisories were somewhat encouraging: Katrina was moving away from the United States, and for the time being, there were no watches or warnings in effect.

But at the hurricane center, the forecasters knew better. Conditions were perfect for major strengthening. Katrina's winds, which had been around 80 miles per hour over Miami, were now up to 100 miles per hour in the southwest Gulf and increasing.

The longer-range computer models had been indicating that Katrina might strike the Florida Panhandle, but as the day progressed, some of the models started indicating a path farther to the west.

Forecaster Stacy R. Stewart was working the day shift. He called his colleagues' attention to the already low pressure in the Gulf, the very low wind shear, and Katrina's improving outflow six miles above the Gulf waters—all signs that major strengthening was likely.

In contrast to the temporary inactivity on the U.S. mainland, there was furious activity in the Gulf of Mexico, with dozens of oil platforms being evacuated. In addition to the threat from the high winds of a hurricane, waves in a hurricane can exceed 50 feet, making it far too dangerous for oil-rig crews to remain at sea.

There are at least two commercial weather companies in the Gulf Coast states that specialize in providing tropical storm and hurricane warnings to the companies that operate offshore oil drilling platforms. These, in turn, work with a fleet of helicopters to get workers quickly off the platforms to safety on the mainland.

Oil prices had been rising through much of 2005, and any disruption in oil drilling and production off the Gulf Coast inevitably would put additional pressure on consumer prices. Not only is Louisiana important because of its proximity to many of the offshore platforms, but much of the oil that the U.S. imports comes through two ocean terminals just south of the mouth of the Mississippi River.

A major hurricane striking Louisiana would have major implications for the United States' energy supply.

While the NHC gathers reports from weather buoys and ships in the Gulf, once a storm has moved across Florida into the Gulf, back over water, NHC redeploys the hurricane hunter aircraft as its primary data source. Much of Friday was spent planning the recon flights, getting the FEMA hurricane team in place, and for some, getting rest before the inevitable next round of warnings had to go back up over the weekend.

Friday morning at 11:00, it still looked like the Florida Panhandle or Mobile, Alabama, would be ground zero for the second landfall of the strengthening Katrina, but that projection was still quite tentative.

FRIDAY EVENING, AUGUST 26

By Friday evening it was starting to look like Katrina's landfall would be farther west.

To make a forecast of maximum accuracy, the computer models that play a key role in forecasting hurricane movement must be fed a wide array of accurate data on the wind, temperature, and pressure structure of the atmosphere in various geographic locations and altitudes. The cliché "garbage in, garbage out" is especially appropriate in hurricane situations. We know, in retrospect, that incomplete data (e.g., a weather balloon launched too late for the data it collects to be included in the model's initial analysis of the weather) has caused major errors in a model's forecast.

Because the data is sparse over the oceans, the NOAA uses a Gulfstream jet to collect key data at higher altitudes than the propeller-driven hurricane hunter planes can fly. And the higher speed of the jet allows it to cover a larger geographic area within the data "window," which is about one hour in duration four times a day. The data collected during each of these windows is essential to measuring the environment surrounding the hurricane. The cost of operating a Gulfstream jet is high, so it's only used in critical situations. The NHC authorized a data-gathering Gulfstream flight for Friday evening.

While computer models are generated four times a day at the NHC, the most important data collection times are midnight and noon London time (by international agreement), which is 8:00 a.m. and p.m. Miami (Eastern Daylight) time. In 2005, computer models made their forecasts in a two-step process.

First, an initial analysis of the world's weather was created. Equivalent to a snapshot, data was collected to discern the location of high- and low-pressure systems, temperature patterns, winds, and so forth. In addition to weather balloons launched from land-based weather stations and commercial aircraft in flight, the propeller-driven hurricane hunter aircraft was gathering data from the ocean surface to about three miles up, and the Gulfstream was gathering data near an altitude of six miles. Once this data was back at the NHC, the hurricane specialist entered Katrina's exact center and intensity data into the model's initial analysis from which the model forecasts would be made.

The models contain mathematical equations that simulate the behavior of the atmosphere. Once the initial state of the atmosphere is described in terms a computer can understand, the models begin the forecast process. The model makes a forecast for six minutes into the future. It then takes that six-minute forecast, treats it as the new initial atmospheric state, and makes another six-minute forecast. The process repeats a third time, and if it was stopped, you would have a forecast for the hurricane's behavior eighteen minutes from the time of the first projection. Instead, the process was repeated over and over until the model created a forecast days into the future.

The models were (and still are) run by the U.S. National Center for Environmental Prediction, the U.S. Navy, Environment Canada, and the European Centre for Medium-Range Predictions. Each of the models treats a bit differently certain equations that describe the atmosphere; this ensures a better overall picture to help guide the human forecaster in creating an accurate forecast.

Even though the models are generated by some of the fastest computers in the world, it takes hours for these marvels of technology to run the data through the forecast programs. Therefore, the forecasts generated from the data gathered in the 8:00 p.m. window would not be available to government, military, and private-sector forecasters until after midnight.

At 11:00 that night, using data from the 2:00 p.m. Eastern-time model runs, the NHC forecasters were even more doubtful about the northward turn earlier forecast, as the storm was stubbornly continuing to move west. The NOAA jet had sent back data indicating that the high-pressure system blocking the northward turn was stronger than previously indicated. The turn toward the Florida Panhandle would be delayed. And the farther west the northern turn occurred, the farther west Katrina would come ashore.

Data doesn't always tell the whole story. Meteorologists often work on instinct. From the very beginning, most of the NHC meteorologists had a bad feeling about Katrina. Based on the newer data, those instincts looked to be right on the money. The computer models were now predicting wind speeds of up to 131 knots or 151 miles per hour: a Category 4 storm.

SATURDAY, AUGUST 27, PRE-DAWN

With the more complete data set from the Gulfstream, the 8:00 p.m. computer model forecasts that became available during the pre-dawn hours Saturday painted an ominous picture: a major hurricane heading straight for New Orleans. At WeatherData, we agreed with the NHC models and started alerting our clients around two o'clock on Saturday morning.

While there are a number of areas where a major hurricane can cause greater-than-normal havoc, New Orleans was well known to meteorologists as the worst case: if a Category 4 or Category 5 hurricane moved across the city, the levees would fail. Thousands would die in the resulting flooding because much of New Orleans is below sea level.

HURRICANE KATRINA DISCUSSION NUMBER 16

NWS TPC/NATIONAL HURRICANE CENTER MIAMI FL

5 AM EDT SAT AUG 27 2005

KATRINA IS LOCATED WITHIN AN ATMO-SPHERIC ENVIRONMENT THAT SEEMINGLY CANNOT GET MUCH MORE CONDUCIVE FOR STRENGTHENING . . .

The headline in the *Times-Picayune* that morning said it all: "Hurricane Center Director Warns New Orleans: This Is Really Scary."

The NWS forecast office in the northern New Orleans suburb of Slidell learned on the nine o'clock hurricane coordination call with NHC that the New Orleans area was being placed under a hurricane watch in an hour. From that point on, the staff was on 12-plus shifts, regardless of family situations. This meant that instead of their normal eight hours on duty with sixteen hours off, they would work twelve on and twelve off until the hurricane passed out of their forecast territory. The impending issuance of a hurricane watch also triggered disaster coordination calls with local officials, including Governors

Haley Barbour (Mississippi) and Kathleen Blanco (Louisiana), and Mayor Ray Nagin of New Orleans.

Throughout the day Saturday, the storm continued to gradually get better organized with higher winds and with winds over a larger geographic area. By late afternoon, to the New Orleans forecasters Katrina looked like "a monster."

At WeatherData, we provided multiple forecast updates for our clients and participated in a number of client conference calls on the increasing threat. We, too, forecast major strengthening of the hurricane.

Among our clientele are a number of insurance companies. Their catastrophe teams were rapidly gearing up based on our forecasts, which were reinforced by what they heard from the NWS. One of our goals is to work with our insurance clients to get their adjusters and support teams into safe positions close to but just *outside of* the area the hurricane is expected to affect. Pre-positioning the insurance teams just outside of the threat area ensures that they can be on the scene quickly to assist their clients as soon as the hurricane wind and rains move out of the area.

SATURDAY AFTERNOON

BULLETIN ... HURRICANE KATRINA ADVISORY NUMBER 18 NWS TPC/NATIONAL HURRICANE CENTER MIAMI FL

4 PM CDT SAT AUG 27 2005

KATRINA RE-ORGANIZING OVER THE SOUTH-EASTERN GULF OF MEXICO ... AT 4 PM CDT ... 2100Z ... A HURRICANE WATCH IS NOW IN EFFECT ALONG THE NORTHERN GULF COAST FROM INTRACOASTAL CITY TO THE ALABAMA-

FLORIDA BORDER. A HURRICANE WARNING
WILL LIKELY BE REQUIRED FOR PORTIONS OF
THE NORTHERN GULF COAST LATER TONIGHT
OR SUNDAY. KATRINA IS A CATEGORY THREE
HURRICANE ON THE SAFFIR-SIMPSON SCALE.
STRENGTHENING IS FORECAST DURING THE
NEXT 24 HOURS . . . AND KATRINA COULD
BECOME A CATEGORY FOUR HURRICANE LATER
TONIGHT OR SUNDAY.

Katrina was forecast to reach New Orleans late Sunday night or early Monday morning.

The hurricane watch included Mobile, the entire Mississippi coast, and much of the Louisiana coast. A mass evacuation of the counties along the central Gulf Coast was needed, as Katrina was not only a potentially severe hurricane, she was unusually large. The hurricane's large size, if it intensified, could cause major damage over a vast geographic area that would stress recovery resources.

In Mississippi, Governor Haley Barbour ordered a mandatory evacuation of Hancock County, the westernmost coastal county and the area predicted to suffer the worst effects of Katrina.

Yet in New Orleans, only a *voluntary* evacuation was in effect, to the utter astonishment of just about everyone in the meteorological community. If there was ever a time to get everyone out, this was it.

SATURDAY, 5:00 P.M.

National Hurricane Center Discussion (a meteorologist-to-meteorologist discussion issued by the NHC):

KATRINA SHOULD STRENGTHEN . . . THE GFDL [a
hurricane forecast model] IS NOW CALLING FOR
A PEAK INTENSITY OF 131 KT . . . WHILE THE

SHIPS MODEL IS NOW CALLING FOR 130 KT AND THE FSU SUPERENSEMBLE [a set of models] 128 KT. THE INTENSITY FORECAST WILL CALL FOR STRENGTHENING TO 125 KT AT LANDFALL . . . AND THERE IS A CHANCE THAT KATRINA COULD BECOME A CATEGORY FIVE HURRICANE BEFORE LANDFALL.

One-hundred twenty-five knots equals 146 miles per hour, a strong Category 4, more than enough for the levees to fail to protect New Orleans and flood the city.

The director of the NHC, Max Mayfield, was so concerned that Katrina would be a major disaster for the Gulf coast and for New Orleans in particular that he took the unusual step of personally telephoning the head of Alabama's emergency management; Haley Barbour, the governor of Mississippi; Kathleen Blanco, the governor of Louisiana; and Ray Nagin, the mayor of New Orleans. The calls were made between 7:25 p.m. and 8:00 p.m. Eastern time, which is between 6:25 p.m. and 7:00 p.m. in New Orleans. Mayfield told each of them that there was the potential for a "large loss of life." Nagin received the call from Mayfield "over the dinner hour."

With the watch that had been in effect all day, plus the personal briefing from Mayfield, one would have thought that Nagin would have immediately gone on local television and radio, announced a mandatory evacuation, and started following his city's and the state's emergency plan for southeast Louisiana. That plan called for the use of city and school buses to get those without transportation out of the city. Unfortunately, none of this occurred Saturday evening. The hours ticked by.

* * *

For years, meteorologists and disaster experts had known that New Orleans was a catastrophe waiting to happen. Because of research done by FEMA, we knew that New Orleans had the third worst potential for loss of lives from a natural disaster. If the city were to remain unevacuated, the winds of a strong hurricane (which would disrupt power and transportation, making last minute escape unlikely) plus the flooding from the failed levees could result in more than 50,000 potential deaths.

These dire statistics were known to many, if not most, New Orleans residents as well. The *Times-Picayune* had prominently featured the scenario of the levees overtopping or breaching in the storm surge of a major hurricane. And because much of New Orleans is below sea level, it would fill up like a bowl once the water started flowing.

With electric power interrupted as a major storm moved through, there would be no way to pump the water out, meaning the water would rise and stay put for days or, more likely, weeks. The city would sit in a soup of water, chemicals, and sewage while the structures underwater would deteriorate. Economic losses would be huge.

But it's one thing to *believe* something and another to *act* on it. While the politicians, bureaucrats, and much of the public knew the levees *could* fail, and that mass casualties might result, they did not seem to be able to envision the city flooding and the resulting death and devastation if they failed to act.

It was not that an emergency plan for responding to hurricane flooding of New Orleans had never been practiced. In July 2004, just thirteen months before Katrina, the "Hurricane Pam" disaster exercise was held. A wide variety of emergency management, disaster agencies, and first responders practiced what they would do if a major hurricane struck New Orleans and flooded the city. The exercise actually *planned* for 60,000 deaths. Officials knew the gravity of the situation in New Orleans: Get everyone out, or else people—in large numbers—will die.

According to a press release from FEMA about the Pam exercise, "We made great progress this week in our preparedness efforts." Well, not quite.

Ivor van Heerden writes about the Pam exercise in *The Storm*: "I got the sense that several federal officials were in attendance only because it was required. Perhaps not everyone at the exercise believed this could actually happen." Van Heerden also notes that during the Pam exercise plans were made to shelter evacuees for ten days. When he asked the Pam organizers what happens on the eleventh day and thereafter to the evacuees (since it was impossible that a flooded New Orleans would be ready for resettlement after just eleven days), he received shrugs for an answer.

SUNDAY

As the day dawned, the Katrina situation had gone from bad to night-marish, as the hurricane had reached Category 4 intensity at 2:00 a.m. Then, things got even worse.

> HURRICANE KATRINA SPECIAL DISCUSSION . . .
> 8AM EDT SUN AUG 28 2005 . . .
>
> THE PURPOSE OF THIS SPECIAL ADVISORY IS TO
> REVISE THE INTENSITY OF KATRINA TO CAT-
> EGORY FIVE . . . MAXIMUM SURFACE WINDS OF
> ABOUT 140 KNOTS (161 MPH).

Katrina was now a Category 5 with 161-mile-per-hour winds, and it was predicted to move across New Orleans.

Forecaster Jack Bevin was under intense pressure because of the implications of his revised forecast—a Category 5 hurricane moving across New Orleans was certain to flood the city, causing a catastrophic loss of life if the population was not evacuated in time.

In New Orleans, the Sunday *Times-Picayune*'s full-color front-page headline was "Katrina Takes Aim: Wall of Water—Levees Could be Topped in the Entire Metro Area." To anyone paying attention, the word was out. Katrina was a critical threat.

I sat at my kitchen table reading this news on my laptop computer that Sunday morning and I remember saying three things to Kathleen: I am going to have to go into the office this afternoon; President Bush needs to get his butt out of Crawford and take personal charge of the recovery effort; and I need to contact a client at a major television network.

The last was the easiest. I sent an e-mail to the client, letting him know that I was concerned this would become the most costly natural disaster in the history of the United States and that the loss of life would be very high because there was *still* no mandatory evacuation in New Orleans. I stated in the e-mail that all the stops should be pulled out in covering this storm.

I had no contacts that could have gotten a message to the president.

Given the size of Katrina (meaning that her winds and storm surge would affect the entire central Gulf coast), and the flooding of New Orleans I was expecting, it would be essential to have the president fully engaged in order to cut through red tape and make sure the recovery effort was fully up to the challenge. Unfortunately, this did not occur. Bush's FEMA director was Michael Brown, who had no real experience or expertise in the field of emergency management or meteorology.

Back in New Orleans, at 10:00 that morning, a mandatory evacuation was finally announced, less than 24 hours prior to the time of predicted landfall. Yet the buses that were part of the emergency plan were not activated. Amtrak's offer to make its resources available to evacuate residents (the Amtrak station is near the Superdome) was turned down. It was also announced that the Superdome would be available for those who could not get out.

Making the Superdome announcement at the same time as the mandatory evacuation announcement was a strategic mistake. Telling people that the Superdome would be available was a huge disincentive for people to get out. Why go to the trouble to leave if there is, nearby, a big, safe structure blessed by the city?

SUNDAY AFTERNOON

Robert Ricks, Jr., of the New Orleans NWS, finished packing the van and sent his family west to San Antonio to ride out the storm. He packed a bag for himself because he would have to be at the weather office for the duration of the storm. He says he took one last look around the neighborhood because he imagined that it would never look the same.

I don't typically go into the office on weekends during hurricanes, since the WeatherData staff is fully capable of handling them. However, with the worst case staring us in the face, I thought my assistance would be helpful to our staff and reassuring to our clients.

When I arrived, the staff had the situation well in hand. Recalling what set our work apart during Andrew, I gave a high-level briefing to the top management of one of our insurance clients. At the end of the conference call, during which I predicted more than a thousand deaths and the flooding of New Orleans, a vice president asked, "Is there *any* good news?" Sadly, I replied, "No." The executives on the other end of the call reacted with stunned silence.

I then set to work on an *Economic and Sociological Effects* forecast for Katrina. We knew this storm would be something extraordinary, and we were concerned that forecasts of the weather, by themselves, would not fully convey the enormity of the situation. We did a short version soon after I arrived. At 5:30 p.m., we released a detailed version.

Our report called for catastrophic damage near the point of landfall; a breach of the levees with severe flooding; serious damage, including widespread power failures and flooding as far east as Pensacola; a loss of life higher than that caused by Andrew and Camille; major loss of life if the eye were to go over New Orleans; total damage possibly exceeding $50 billion; and significant damage to Loop Terminal, Port Fourchon, and the Port of Mobile (Loop and Fourchon are the terminals through which much of the United States' imported oil passes).

Our clients pay us for our best estimate of what will happen. As I typed the words in the report, even I was having trouble believing them. Selfishly, I feared I might lose face if, somehow, Katrina suddenly weakened or if the levees miraculously held.

Ricks, back at the National Weather Service office in New Orleans, also did an outstanding job writing strongly worded statements for the general public that forecast "incredible human suffering" and predicted the area would be "uninhabitable for weeks."

It turns out Ricks had similar doubts after he sent his message. It was so strongly worded that he later wrote, "calls began to come in from the national media inquiring about the legitimacy of the warning. I confirmed that I issued it and that it was authentic . . . Still, after all of the warnings had been issued, I began to worry that I was the forecaster who had cried wolf. As much as the meteorologist in me wanted to be right, the concerned human in me wanted to be very wrong."

* * *

The success story of the warnings: One million people were evacuated from New Orleans and the central Gulf coast. When they left their homes, many of them expected to have to drive only fifty or so miles. Because there were not enough nearby hotel rooms to handle the exodus, some evacuees ended up as far away as Arkansas and Huntsville, Alabama. Most of them took only enough clothing and provisions to last a day or two.

There was little to do now but wait.

08/28/05 2245Z GOES-12 VIS

Hurricane Katrina a few hours before landfall.

CHAPTER TWENTY-ONE

KATRINA: PART THREE— MURDER BY BUREAUCRACY

Bureaucracy has committed murder here in the greater New Orleans area.

—Aaron Broussard, *Meet the Press*, September 4, 2005

MONDAY MORNING, AUGUST 29

WHEN I ARRIVED AT THE OFFICE AROUND 5:00 A.M., Katrina had weakened to Category 4 and was just off the coast of Louisiana. It was a few miles east of our predicted location, but our forecast at this point was excellent. The eye of Katrina came ashore at 6:10 a.m. near Buras, Louisiana, about as we had expected.

Over the next hour and a half the storm started to draw in dry air that caused it to weaken faster than predicted. As Katrina weakened, she took a slight turn to the right, which carried the center just east of New Orleans. This is critical, as the strongest winds in a hurricane are in the eye wall (the ring of clouds around the nearly calm center of the storm) and to the right (east) of the eye. While damaging wind gusts of up to 120 miles per hour occurred in eastern New Orleans and caused extensive roof damage, Category 4 and Category 5 winds *never* occurred in New Orleans during Katrina.

Shortly after passing east of New Orleans, the eye of Katrina made its final landfall as a weak Category 3 on the Mississippi-Louisiana border. When it made landfall, the howling winds along the Mississippi coast were almost deafening, but it wasn't the wind that Mississippians will remember when they think of Katrina.

It's important to understand that a storm surge develops in hurricanes due to a combination of low air pressure (which, in apparent defiance of gravity, actually raises sea level in and near the eye), high winds, and the shape of the coastline. It has long been known that the biggest killer in hurricanes is water, not wind. And the storm surge from Katrina in Mississippi may have been unprecedented, especially for a Category 3 storm.

While Katrina was offshore at Category 5 intensity on Sunday, a tremendous amount of water built up and her winds generated waves that may have topped at 90 feet. NOAA weather buoys offshore have never before (or since) recorded wave heights so high. A giant and supposedly indestructible oil platform was knocked out of service, and for the first time ever, a NOAA weather buoy was destroyed by Katrina's wind and waves. When the hurricane's eye arrived at the coast, the water and waves—churned up by wind gusts of nearly 200 miles per hour offshore—arrived with such force they took the Mississippians who stayed behind by surprise.

Some Mississippi coastal residents didn't evacuate in spite of the dire hurricane warnings and mandatory evacuation: People who survived Camille thirty-six years before thought they had seen the worst that Mother Nature could do ("it was a Category 5!"). At least a few of those people, based on anecdotal evidence, were among the 238 who perished in Mississippi. They had learned the wrong lesson from Camille: that you can remain on the coast in a Category 5 and survive.

Shortly after sunrise, the storm surge reached the Mississippi coast and drowned everything in its path with incredible force. Reaching a depth of 28 feet, the surge conquered the coast to at least six miles inland (crossing Interstate 10 in many locations) and as far as twelve miles inland along bays and rivers. Entire towns and neighborhoods were flattened. Every single square foot of Mississippi under the 28-foot surge had more than 1,700 pounds of water pushing down on it. Put another way, a 10-by-10-foot room had 170,000 pounds—85 tons—of pressure on it. Obviously, few survived.

Storm-chaser Jim Reed's photographs of an automobile surging through the lobby of a hotel give just a taste of the surge's force, and his photographs of the aftermath give testament to the sheer force of water (see color insert).

The Louisiana-Mississippi border area is where the NWS made the one major misstep in its otherwise exemplary performance: At 7:25 a.m. CST, the NWS began putting out tornado warnings for Katrina's eye wall winds. The tornado warnings ended with the words, "Seek shelter on the lowest floor of the building in an interior hallway or room such as a closet. Stay away from windows and remain in your safe shelter until the eye wall passes."

As Reed's photos illustrate, there could hardly be a less safe place than the lowest floor in the face of a 28-foot storm surge. Whether anyone actually followed the NWS's unfortunate advice is unknown.

Back at WeatherData, we were monitoring the storm and providing updates to our clients. The pace was busy but doable. At 8:12 a.m., the New Orleans NWS office's hydrologist Patricia Brown received a radio transmission from Bob Turner, the Lake Borgne levee district manager, that the Industrial Canal levee was breached on the east side at Tennessee Street.

Two minutes later, the NWS issued a flash-flood warning:

A LEVEE BREACH OCCURRED ALONG THE
INDUSTRIAL CANAL AT TENNESSEE STREET. 3
TO 8 FEET OF WATER IS EXPECTED DUE TO THE
BREACH. LOCATIONS IN THE WARNING INCLUDE
BUT ARE NOT LIMITED TO ARABI AND THE 9TH
WARD OF NEW ORLEANS. IF YOU ARE IN THE
WARNING AREA MOVE TO HIGHER GROUND
IMMEDIATELY.

As we read the warning as it printed out, we knew this was it: the worst was going to occur. We got on the phones, started telling our clients that New Orleans was flooding due to a levee breach, and prepared written updates with the news.

At 2:00 that afternoon, New Orleans City Hall confirmed the breach and said 20 percent of the city was flooded. The flood was slowly and inexorably spreading. There was still time, however, to mount an emergency effort to get people out since the dangerous winds from Katrina had subsided.

The receipt of the 8:14 a.m. NWS flood warning was met by the media and emergency management in New Orleans and the surrounding area with the following reaction: Nothing. By late morning, some of the TV networks were reporting that New Orleans had "dodged a bullet," even as the city was filling with water.

Ivor van Heerden, in *The Storm*, describes the mood in the Louisiana State Emergency Operations Center (EOC) in Baton Rouge, where he spent the day as an expert in storm surge flooding, as "back-slapping" at 8:00 p.m. that night. Van Heerden, who spends many of the pages of his book blaming President Bush for the disaster, excuses himself, local emergency officials, and the state emergency operation by saying the NWS's flash-flood warning got "lost." The warning was sent through multiple communications channels. How lost could it have been in Baton Rouge if we received it immediately in Wichita?

The 8:00 p.m. celebration time is important because that is when van Heerden and the others at the EOC received a report of two feet of water in a New Orleans nursing home. How does van Heerden react to the report? In his own words, he goes home (although he later returned).

The amount of incompetence at all levels of government, local, state, and federal, up to this point was breathtaking; and it was just getting started.

* * *

I had a business trip to Raleigh, North Carolina, planned for that afternoon to train a new client on our storm-warning system. I debated whether to go, but since things were calming down, and my staff was on top of things, I decided to make the trip.

I checked into a hotel near the client's facility and immediately turned on the television to find out the latest on the New Orleans flood. Nothing.

Unbelieving, I went to a late dinner and returned to the room. Still nothing. I stayed up to watch *Nightline* on ABC, thinking that surely *they* would have accurate information, as it had been more than 15 hours since the levee breach had occurred.

Instead, Chris Bury was grilling David Johnson, the head of the NWS, about how the service had *overforecast* Katrina. Knowing Chris to be a competent reporter (we had worked together in St. Louis), I was extremely disappointed at the sophomoric line of questioning. The entire interview cast the NWS in the role of the "boy who cried wolf." I'm sure that as he prepared for the interview, David expected well-deserved accolades from Chris; instead, he received brickbats. I felt sorry for David and thought he handled himself wonderfully. Even though it was almost midnight after a very, very long day, I immediately composed an e-mail to David telling him he represented the entire profession well during the interview and congratulating him on the NWS's performance during Katrina.

TUESDAY THROUGH SUNDAY

By Tuesday, 80 percent of New Orleans was flooded due to numerous breaches of the levees.

The next few days live in infamy as every level of government screwed up beyond belief. The city of New Orleans filled with water. Citizens who survived the hurricane died, by the hundreds, in the post-storm flood. People were stranded at the Superdome and convention center, the latter of which wasn't even supposed to be a shelter.

The cavalry arrived in the form of Wal-Mart, the Red Cross, the Salvation Army, and insurance companies, but unbelievably, they were turned away by law enforcement. New Orleans storm survivors walking across the bridge to nonflooded Gretna were turned away at gunpoint by the state police, their fellow Louisianans.

And finally, trained firefighters and other first responders urgently summoned from across the country to help get people to safety were first flown to Atlanta. Why Atlanta? Why not, say, Houston, Baton Rouge, Fort Polk—in other words, someplace close?

There was a reason for Atlanta. While people were dying by the hundreds, the federal government forced these brave rescuers to take a course in sexual harassment *prior* to being deployed.

There is no question the out-of-town rescuers' work environment in New Orleans was hostile. But it had nothing to do with sexual harassment. The hostile environment was due to flood, fire, sewage, snakes, lack of electricity, armed looters (which included a few members of the New Orleans police department), and dozens of other problems and challenges. The pathetic requirement to sit through a sexual harassment course was an insult to the brave men and women who volunteered to help their fellow citizens in need. It also says a lot about the priorities of the federal government, especially since, as far as I have been able to determine, no one has been fired or otherwise called to account for this unbelievably stupid decision.

Van Heerden goes to great lengths in *The Storm* to indict the Bush administration for its sloppy handling of Katrina, and justifiably so. President Bush's flying over New Orleans in *Air Force One* on his way to D.C. from Crawford didn't cut it. Neither did FEMA director Michael Brown's concern about what he was going to wear for TV interviews.

Brown's convoluted priorities are revealed in this telling e-mail exchange between Brown's senior staff members: August 31, 11:20 a.m. e-mail from Marty Bahamonde regarding the scene in downtown New Orleans, to Michael Brown in Baton Rouge:

> Sir, I know you know the situation is past critical. Here are some things you might not know. Hotels are kicking people out, thousands gathering in the streets with no food or water . . . We are out of food and running out of water at the [Super]dome."

2:00 p.m. e-mail from Sharon Worthy (Brown's press secretary) to Cindy Taylor, the FEMA deputy director of public affairs, and other FEMA staffers:

> Also, it is very important that time is allowed for Mr. Brown to eat dinner. Given that Baton Rouge is getting back to normal, restaurants are getting busy. He needs much more than 20 or 30 minutes. We now have traffic to encounter to get to and from a location of his choice, followed by wait service from the restaurant staff, eating, etc.

2:44 p.m. e-mail from Bahamonde to Taylor and Michael Widomski at FEMA public affairs:

> OH MY GOD!!!!!!!! . . . Just tell her that I just ate an MRE [a canned military-issued "meals ready to eat" intended for

use in combat] and crapped in the hallway of the Superdome along with 30,000 other close friends so I understand her concern about busy restaurants.

But it wasn't just the Bush administration. Congress was equally useless. Mississippi Senator Trent Lott, while safely in Washington, D.C., whined about his insurance company's alleged slow service in restoring his luxury coastal home. Louisiana's Senator Mary Landreau spouted, "I'm going to literally hit George Bush!" because she judged the administration's response as inadequate.

The locals were even worse. The governor of Louisiana, Kathleen Blanco, seemed paralyzed by indecision throughout the crisis.

> Roundup of buses for storm bungled: Blanco documents show staff confusion (*Times-Picayune*, Tuesday, December 06, 2005) by Laura Maggi
>
> BATON ROUGE—Two days after Hurricane Katrina hit New Orleans, thousands of people were trapped in the city without food, water, and medical care, and growing increasingly desperate for rescue. But a top aide to Gov. Kathleen Blanco sent out an e-mail informing his colleagues that his staff had stopped calling for the buses needed to evacuate people from the Superdome and other places of refuge.

In a flooding city, the Blanco administration didn't want buses to get people out.

A little realized fact about the Superdome: Even though Katrina's winds never reached anywhere near Category 4 or Category 5 strength, the roof was significantly damaged during Katrina, to the extent a hole opened up in it. Had winds of that strength actually occurred in downtown New Orleans, the Superdome's roof might have completely failed with catastrophic results for those taking shelter inside.

Finally, there's Mayor Ray Nagin.

Nagin still doesn't get it. During Mardi Gras 2006, he told Katie Couric that he "wished Max Mayfield had called him sooner." But it wasn't NWS director Mayfield's job to hold his hand during the run-up to Katrina. It was Nagin's job to follow the city's hurricane emergency plan to declare a timely mandatory evacuation and use the buses to get the people who couldn't fend for themselves out of the city. And if Nagin had acted immediately upon receiving Mayfield's "going the extra mile" phone call, the mandatory evacuation could have started more than twelve hours earlier than it actually did.

And finally, while Nagin was frequently critical of Bush, Brown, and the rest of the federal response, where was Nagin? At the Hyatt Regency New Orleans. The Hyatt was without power and had windows blown out in the storm, so it was not the very pleasant hotel it normally is. But there is an enclosed elevated walkway from the second floor of the hotel right into the Superdome. Nagin could have visited his constituents and supervised the crisis there without even getting his feet wet. He never did so, as far as I can determine.

In fact, while the city was still flooded and his citizens were still suffering, on September 4 Nagin called for his police to be sent on taxpayer-paid rest-and-recreation trips to Las Vegas!

Contrast this with the work of the New Orleans' NWS office in Slidell. After 12 plus hour shifts in the run-up to the storm, a number of employees lost their homes and had to sleep in the office for *weeks* while continuing to do their duties. They viewed the inconvenience as part of their duty to their fellow citizens.

Besides the great work and forecasting of the NWS, about the only good thing I can point to in terms of a positive, proactive government agency response was the Coast Guard's heroic rescue efforts to pull people from the roofs of flooded New Orleans structures, for which they deserve tremendous praise.

The purpose of this book is to tell the story of how the science of weather has improved American life. The meteorological profession (both the NWS and private-sector weather industry) performed admirably during Katrina. Unquestionably, thousands—and likely tens of thousands—of lives were saved. It is beyond upsetting to see that legacy destroyed when people who survived the hurricane died in the anticipated and planned-for flood after the levees were breached.

KATRINA'S AFTERMATH

The most precise estimate of the death toll from Katrina is 1,833 over five states: Louisiana (1,577), Mississippi (238), Florida (14), Georgia (2), and Alabama (2), as of August 10, 2006, with "several hundred" reported missing, some of whom undoubtedly washed out to sea in the storm surge; their fate will likely never be known.

With that qualification, it's safe to assume that more than 2,000 people actually perished in Katrina. Still, the death toll was not remotely close to the 50,000 or more (predicted by multiple studies) expected in a hurricane-caused levee breach. Why? The timely warnings that allowed a million people to evacuate. Meteorology did its job.

From the National Hurricane Center's final report on Katrina:

> The American Insurance Services Group (AISG) estimates that Katrina is responsible for $40.6 billion of insured losses in the United States. A preliminary estimate of the total damage cost of Katrina in the United States is assumed to be roughly twice the insured losses, or about $81 billion. This figure makes Katrina far and away the costliest hurricane in United States history. Even after adjusting for inflation, the estimated total damage cost of Katrina is roughly double that of Hurricane Andrew (1992). Normalizing for inflation and for increases in population and wealth, only the 1926 hurricane that struck southern Florida surpasses Katrina in terms of damage cost. The Insurance Information Institute reports that, mostly due to Katrina but combined with significant

impacts from the other hurricanes striking the United States
this year, 2005 was by a large margin the costliest year ever
for insured catastrophe losses in this country.

But here is the sobering bottom line: Katrina turned out *not* to be
the worst case. She weakened before she hit land. Had she maintained
Category 5 strength, the flooding in New Orleans would have been
quicker (due to additional breaches), likely resulting in more deaths.
The impressive storm surge in Mississippi, Alabama, and Louisiana
would have been higher (also resulting in more deaths), and the inland
wind damage would have been worse.

New York City has been hit by hurricanes and will be again. The
Baltimore-Washington area is vulnerable if a major hurricane moves
up Chesapeake Bay. Andrew did not hit the heart of Miami; a direct
hit will be far worse. Houston-Galveston is a huge, growing, and
vulnerable population center. And there's one area few people have
considered: while rare, hurricanes *have* and do hit California. There
are zero plans in place for when a hurricane returns to the West
Coast. We must not become complacent: there are many possible
"Katrinas" lurking.

Finally, there needs to be a culture change in Washington. I was
chair of the Commercial Weather Association's lobbying committee
for a number of years, so I got to see, close up, how the process works.
The insiders say, "in Washington, nothing succeeds like failure." What
they mean is that it is difficult to obtain funds and support for proactive, constructive measures. But when there is a major governmental
failure, such as the response to Katrina, the money flows, and equally
important from a cultural standpoint, nobody gets fired.

In the case of Katrina, Nagin was reelected. Blanco continued
to serve out her term. The staffers who were worried about where
Michael Brown was going to eat continued in their jobs. Yes, Michael
Brown was temporarily "fired" as director of FEMA, but he was

rehired by FEMA as a high-paid consultant. President Bush took a big hit in popularity, but otherwise no one was called to account.

The responses to both Andrew and Katrina were complete fiascoes. The Department of Homeland Security seems to believe that its primary mission is harassing grandmothers wanting to board airplanes rather than providing an *effective* airport security or effective responses to major disasters. It is probably unrealistic to expect government, especially in a democracy, to operate as efficiently as private business.

That said, the current bar is awfully low. We must demand more efficiency and accountability from those we send to represent us in our cities, our state capitals, and in Washington.

CHAPTER TWENTY-TWO

GREENSBURG: CAPSTONE OF THE MODERN WARNING SYSTEM

EARLY IN THE EVENING OF MAY 4, 2007, LANCE Ferguson and his wife, Cyndi, were having dinner parked in their truck near the World's Largest Hand-Dug Well, in the small town of Greensburg, Kansas. They later came to refer to it as "the last supper" because, in just a few hours, Greensburg would cease to exist.

Greensburg was a tree-filled, quirky little town enjoyed by Midwesterners whenever they passed through on U.S. Highway 54, the primary east-west highway in the southern half of the state. Almost everyone in Kansas seemed to have visited the Well at one time or another.

But it wasn't just the Big Well and the trees that contributed to the town's charm. A huge meteor was on display next to the Well. Greensburg boasted a true western store patronized by bona fide cowboys. The Hunter Rexall Drug Store, a community gathering place, still had an old-fashioned soda fountain, with an eighty-two-year-old soda jerk, who had held the job since 1952; he was making cherry Cokes and scooping ice cream for sodas while the Fergusons were enjoying their dinner.

Lance Ferguson, then a real estate agent in his full-time profession, is a storm chaser for Wichita's KWCH-TV. He'd cleared his calendar four days earlier, when weather computer models indicated a relatively high probability of severe thunderstorms, possibly accompanied by tornadoes, in Kansas on Friday and again over the weekend.

When Friday morning rolled around, Ferguson started his day by putting his chase gear in the trunk of his car. He kept an eye on the preliminary forecast information issued by NOAA's Storm Prediction Center in Norman. KWCH sent an e-mail to its team of chasers indicating that they would likely be needed later in the day.

Over the noon hour, a very weak front moved south through the southern half of Kansas and then became more or less stationary the rest of the day in the vicinity of Highway 54. There was little temperature difference on either side of the front; it was just a wind shift from southeast to east-southeast.

By early afternoon, after monitoring the weather system all day, I was concerned that the situation was evolving into one conducive for violent tornadoes. The only question I had was whether a stable layer of the middle atmosphere three miles above the ground over southwest Kansas was pronounced enough that it would stop the thunderstorms from developing. Called "the cap" by meteorologists because the stable layer caps the rising air and chokes off the development of thunderstorms, it can be one of the most tricky things to forecast in severe weather situations. If the cap held, there would be no thunderstorms. Nevertheless, I was concerned enough to walk back to our IT department to make sure that all of the primary and backup systems at WeatherData were running properly in case the cap "broke." If so, intense thunderstorms would rapidly develop. I didn't want to go into this particular situation with any equipment problems. After being assured that everything was running fine, I invited myself to our client station, KWCH, which serves all of central and western Kansas. We had just delivered some new software for KWCH's severe storm analysis and display system, and I wanted to make sure people there knew how to use it; I was convinced they were going to need it.

I arrived at KWCH around 4:10 p.m. and went over the software's new capabilities with Merril Teller, the chief meteorologist who was on duty that day. Joining us was meteorologist Lindsay Boutor, who asked me what I thought of the tornado potential. I told her it was very high if the cap were to break. I stayed at KWCH through the five o'clock newscast. About the time the weathercast hit the air at 5:15, thunderstorms were starting to develop near the Texas Panhandle–Oklahoma border.

Shortly after I left KWCH, the Storm Prediction Center in Norman issued Tornado Watch 227:

URGENT . . . IMMEDIATE BROADCAST REQUESTED . . . TORNADO WATCH NUMBER 227

NWS STORM PREDICTION CENTER NORMAN OK . . . 615 PM CDT FRI MAY 4 2007

THE NWS STORM PREDICTION CENTER HAS ISSUED A TORNADO WATCH FOR PORTIONS OF WESTERN AND CENTRAL KANSAS . . . WESTERN AND NORTHERN OKLAHOMA . . . THE EASTERN TEXAS PANHANDLE AND NORTHWEST TEXAS EFFECTIVE THIS FRIDAY NIGHT AND SATURDAY MORNING FROM 615 PM UNTIL 200 AM CDT

TORNADOES . . . HAIL TO 3 INCHES IN DIAM-ETER . . . THUNDERSTORM WIND GUSTS TO 70

MPH . . . AND DANGEROUS LIGHTNING ARE POS-
SIBLE IN THESE AREAS.

It looked like southwest Kansas was going to be the area where
the tornado potential was highest. Several hours earlier, the KWCH
chasers had headed west: Eric Lynn to Dodge City, Scott Roberts to
Kinsley, and the Fergusons to Greensburg.

They weren't alone. Since the purely scientific chases of the 1970s,
storm chasing has become a cottage industry in the Great Plains in
spring. You can even book a two-week storm-chasing vacation from
Tempest Tours, Silver Lining Tours, and others. At least one chaser
had flown to Dallas from London and had driven to Kansas for that
evening's storms. There's even a website where users can track chasers, using GPS technology.

By early afternoon, there was a mass exodus of chase vehicles westbound from Wichita on U.S. 54, and westbound from Oklahoma City
on Interstate 40.

Few things are more spectacular than a powerful Kansas thunderstorm. When observed safely, it's a feast for the eyes, ears, nose
(rain-saturated air smells great), and even touch (rapid changes in
temperature and gusting wind feel amazing). But storm chasing done
incorrectly can be extremely dangerous, even fatal, as the events of the
next few hours would demonstrate.

Scott Roberts reached Kinsley about 3:30 p.m. He found a Wi-Fi
hotspot and sat there for about 90 minutes analyzing the weather.
He was there so long a police officer stopped and asked what he was
doing. Roberts briefed the officer on the likelihood of severe storms
in the area. Roberts noticed that a weak front was indicated on the
Dodge City radar. Based on his experience, it appeared to him that
the likelihood of tornadoes was rapidly increasing farther south near
Coldwater. He turned the ignition key.

While I was doing last-minute training at the KWCH studios, not
much was happening in Greensburg other than a few cumulus clouds

dotting the sky. For a time, it looked like the cap might hold, suppressing thunderstorms in southwest Kansas.

Ferguson even thought about returning to Wichita but decided to stay in Greensburg. The couple got some sandwiches and sat under a large shade tree at the Well near the south edge of town, where Ferguson used wireless Internet to update himself on the developing weather situation. There were a couple of thunderstorms near the Oklahoma-Texas border about 80 miles away that were moving generally north, toward Greensburg, but they were hours away.

Ferguson viewed the cells coming north from Oklahoma with alarm. The Fergusons finished their food and headed south on U.S. 183. They considered traveling toward the strongest storms, still in Oklahoma, but decided to stay in Kansas so they would be in KWCH's coverage area to help warn its viewers. When they got to the town of Protection, Cyndi Ferguson got behind the wheel so her husband could concentrate on storm spotting.

At WeatherData, our team of meteorologists was also watching the Oklahoma and Texas thunderstorms closely, since we cover three major rail lines in the central high plains. At 6:21 in the evening, a small tornado developed near Arnett, Oklahoma. We had already issued a tornado warning: the trains were safe, and nothing was out of the ordinary so far.

About 6:45, as the thunderstorms approached the Kansas border, it looked like they might weaken, as the radar showed intensity levels decreasing. That was the influence of the cap, which, if strong enough, would snuff out thunderstorms. If one only had radar on which to base a judgment, it looked like the invisible cap might hold.

The totality of data told a different story. Shortly after 5:00 p.m., barometers measured rapidly falling pressures. In the earlier chapter about Woodward, I mentioned the lifted index and that a reading of minus eight or lower was practically off the chart. The air flowing toward Greensburg had a lifted index of minus eleven, one of the lowest I'd ever seen. If the cap didn't hold, all hell would break loose, since the lifted index is a measure of the atmospheric energy available to feed the storm.

At the town of Haviland, ten miles east of Greensburg on U.S. Highway 54, NOAA has an upward-pointing Doppler radar known as a wind profiler. That data told us that the state of the atmosphere was changing rapidly over southwest Kansas. Wind speeds six miles above the ground started increasing rapidly, reaching 100 miles per hour. Those winds brought slightly cooler air to the middle atmosphere over southwest Kansas, eroding the cap.

Once the cap was gone, it was only a matter of time.

A few minutes after seven o'clock, the Dodge City NWS NEXRAD showed the first thunderstorm crossing the border from Oklahoma into Kansas. Located on a small plateau a few miles northeast of Boot Hill, the radar and the Dodge City NWS office under it had responsibility for warning Kiowa and surrounding counties in case dangerous weather approached.

NWS meteorologist Mike Umsheid was watching the approaching storms with an increasing sense of apprehension. The brief weakening trend had ended and the storms were starting to strengthen.

The wind vane at the Dodge City NWS office most of the afternoon indicated the winds were blowing in from the southwest, behind what is called a dry line, the boundary separating dry air from New Mexico from the very moist minus-eleven lifted-index air just a few miles to the east.

By evening, the weak front that had passed through Wichita earlier in the day was still stationary just south of Highway 54. Southeast winds were bringing muggy air in from the Gulf of Mexico near the intersection of the two fronts. If you plotted each weather station's data and drew arrows along the flow of the wind, you would see that the air was flowing in a converging pattern, with Greensburg at the bull's-eye.

Converging winds enhance supercell development. Combined with the extreme atmospheric instability, any thunderstorm that might develop in that area would be very, very severe with the potential for giant hail and exceptionally violent tornadoes.

Scott Roberts sat on a high hill south of the Ferguson's location near the junction of Highways 160/183 and K-1, choosing that spot because of its being at the edge of cellular telephone coverage in the area. Shortly after 7:30 p.m., he noticed a rapidly developing thunderstorm with obvious signs of rotation. It developed so quickly it wasn't even showing up on radar. He doubled back to Protection so he could upload a brief video clip to KWCH to demonstrate how serious the situation was becoming.

From 7:30 to about 8 o'clock, several small but very strong thunderstorms were in Clark and Comanche Counties. These produced funnel clouds and tornadoes in open country. About 8:00 p.m. a new thunderstorm developed and, in a word, exploded. Usually thunderstorms are visible on radar for tens of minutes before they begin to rotate; this one wasn't behaving that way at all. As the sun touched the horizon, Roberts saw its first funnel cloud, a small whirl. A few minutes later, at 8:34, Roberts saw the circulation that eventually developed into the Greensburg tornado. The storm was, indeed, exploding, and doing so just as darkness was starting to envelop the area.

At WeatherData, the situation was also viewed with alarm. At 9:00 p.m., rapidly developing already-severe storms were near Hays and southeast of Kinsley. More ominous was a cluster of large, rapidly intensifying storms extending from Highway 54 at Greensburg south to Highway 160 east of Coldwater. And these storms were intensifying as the last twilight disappeared. All but one of the storms started moving rapidly northeast. The final storm, towering more than 60,000 feet into the atmosphere, started lumbering north as it ingested high-energy air. Violent tornadoes are rare after dark because the energy needed to sustain them usually lessens after sunset. But given the minus-eleven lifted index, the atmosphere east of Dodge City had all of the energy it would need—and more.

At this point, the situation was transitioning from a typical tornado threat to one that could be a mega-disaster. It was dark. The atmosphere was primed to produce large and violent tornadoes. And in the darkness, it was dangerous for the spotters to be out observing the storms. A decade earlier, some or all of the chasers would have gone home due to the danger of being overtaken by a tornado in the darkness, potentially allowing a tornado to go unreported. This night, technology would make a crucial difference.

Cell phones make reporting a tornado easier, but it's still not as simple as you might imagine. You still have to make a connection with someone at the other end. On May 8, 2003, I intercepted an F3 tornado near Lyndon, Kansas. As I drove to where I thought the tornado would develop, I was listening to a nearby radio station, WIBW, in Topeka. It was obvious the broadcasters were not aware a tornado was on the ground to their south. I called the NWS in Topeka a couple of times to let them know a large tornado was on the ground; the line was busy both times. I tried to call 911 in Osage County and couldn't get through. I called WIBW. The line rang and rang, but no one answered. Here was a dangerous tornado and I could not get through to anyone who could warn the public. I knew there had to be a better way.

So I designed an additional capability into Storm Hawk: *Reporter*. With just a few taps of the stylus on the Storm Hawk screen, a chaser, law enforcement officer, or emergency manager could report a storm (existence and location of a tornado, large hail, high winds, etc.) to the NWS, emergency management, or a TV station with GPS precision. No dialing. No busy signal. No diverting meteorologists from other critical duties, such as typing a warning or watching the radar, to take a phone call. Plus, the report automatically displays on a SmartWarn screen at a TV station or at an emergency operations center, with the location of the storm in proper relation to the radar, so quick and more accurate warning decisions can be made.

At 9:00 that night, Scott Roberts sent a Storm Hawk report of a funnel cloud just east of Mount Lookout, an area of high hills northwest of the town of Protection. Roberts reported it, then briefed a couple of law enforcement officials who had pulled up near his vehicle.

At WeatherData and at KWCH, the funnel cloud report instantly appeared on the on-air SelectWarn screen, informing their meteorologists and viewers of the increasing threat. The storm was moving through ranch country but was heading toward our client, Union Pacific Railroad, at Greensburg. The railroads normally want about twenty minutes of warning to get their trains stopped (or, if there is time, out of the area). WeatherData meteorologists Scott Breit, Andrew Crouch, and Trevor LaVoie briefly conferred and decided this situation was so dangerous that they were going to give Union Pacific even more advance notice than usual.

At 9:12 p.m., they issued a tornado warning for Union Pacific from Milepost 326 (nine miles west of Greensburg) to Milepost 342 (seven miles east of Greensburg) valid from 9:30 p.m. until 10:00 p.m. Once the warning was composed, the meteorologist clicked the "send" dialogue box on his screen.

In downtown Omaha, at Union Pacific's Harriman Dispatch Center, the screen of the dispatcher controlling the Golden State Route, Darron Rohe, instantly turned red. Working the "Dispatcher 72" position, he tapped the "acknowledge receipt" button that sent a message back to WeatherData confirming the critical warning had been received.

Then Rohe went to work. Two freight trains were speeding toward Greensburg. He radioed the crews of both trains and told them to stop at the east and west edges of the warning.

> **Dispatcher:** "Dispatcher 72 to the Union Pacific 4307 West, over." (Trains are called by the unit number of the lead locomotive and direction of travel.)

Train's conductor: UP 4307 West, over.

Dispatcher: Guys, don't get in too big a hurry to go westbound. I have a tornado warning, ah, between 325 to milepost 342 until 2000 [10pm], so for about another 35 minutes. Expect a delay at Wellsford.

The trains eventually slowed to a stop at the edges of the warned segment of track.

At 9:19 p.m., Mike Umsheid at the Dodge City NWS issued a tornado warning for Kiowa County, including Greensburg. The Wichita TV stations, which had already been following the storm, interrupted programming and conveyed the warning.

Greensburg's outdoor tornado siren screamed the warning to anyone in town not watching television or listening to the radio. The standard procedure during tornado warnings was to sound the siren for five minutes, then turn it off. Running it longer risked burning it out, with a replacement cost of many thousands of dollars. Former Sheriff Ray Stegman, serving as a storm spotter that evening and concerned about the reports he was hearing of the tornado to his south, radioed the police dispatcher and said, "Leave it on." The siren continued to wail until the power failed.

Back in Dodge City, lightning was flashing as a thunderstorm moved right over the radar. Unlike Udall, fifty-two years before, when the radar couldn't penetrate the thunderstorms causing the storm to falsely appear to weaken, the WSR-88D was powerful enough to accurately depict the storm in Kiowa County in spite of the storm over the radar. As the chaser reports of the tornado's position kept coming in, Mike Umsheid could hardly believe his eyes as he watched his radar. The circulation kept getting stronger and stronger. One hundred miles per hour of circulation. One hundred twenty on the next image. The tornado just kept getting bigger and more powerful.

Looking north from the Kiowa-Comanche county line at 9:30, the tornado no longer looked like the typical funnel. Each flash of lightning revealed a growing cloud mass dragging along the ground. Scott Roberts was following the tornado from behind, staying south of the storm as it moved north-northeast. Roberts was on the air with Merril, giving an audio blow-by-blow report as the SelectWarn showed his Storm Hawk moving north along Highway 183. Merril was describing the radar and giving his viewers the tornado safety rules so they would know how to protect themselves.

At 9:38, there was a large flash that Roberts knew to be an electrical transformer exploding. The tornado had reached an electrical transmission line running north-south along the west side of Highway 183. The tornado's twisting winds grabbed the power lines with such force that dozens of power poles to the south of it were uprooted or snapped and thrown east across the highway, while the poles to the north of the tornado's core were thrown west—in the opposite direction. Power was lost over part of the area.

Roberts' Storm Hawk symbol on the KWCH-TV SelectWarn stopped moving at 9:43 as his advance was blocked by the power lines and debris across the road. On air, he reported that farmhouses south of town had been destroyed.

While Roberts was blocked, the Fergusons were heading north along a country road a few miles west that paralleled U.S. 183. They, too, could see the huge tornado as it headed north-northeast about seven miles ahead and to their right.

In Greensburg it was raining and hailing as the tornado siren continued to scream. The hail reached golf ball–size at 9:48. Most residents, having heard the TV blow-by-blow reports of uprooted trees, downed power lines, and shattered farmhouses to the south, were either in or were heading into their basements when the power failed. If they didn't have a basement, they were moving into closets and bathtubs.

Except for along Highway 54, traffic was nearly nonexistent. Because Highway 54 is a major east-west artery, it carries truck and automobile traffic from all over the country. As people from outside the Midwest were driving through Greensburg, they wondered what the wailing siren and near-deserted side streets were all about.

Along the Union Pacific, the two trains on either side of Greensburg were stopped, and then two more behind them stopped. The four trains would remain so until the warning expired. Back on KWCH-TV, Merril was telling viewers about the large tornado that was on the ground. He also took live calls from Roberts and then Ferguson, who kept emphasizing the size of the tornado and how Greensburg was in mortal danger.

Merril noted that the tornado appeared to be headed toward the east side of the town of population 1,500. Studies have shown that multiple sources of information add credibility and help convince people to take action. So the use of radar and SelectWarn, and Merril's expertise, along with Roberts' and Ferguson's reports from the field, provided the level of credibility and authority to convince people to take the action needed to save their lives.

In the final minutes as the tornado bore down on Greensburg, Mike Umsheid watched as the rotational wind shear increased to more than 260 miles per hour, the highest ever measured on any of the WSR-88Ds.

The tornado, which had held steady on a north-northeast course, took a left turn just south of town. The tornado was 1.7 miles wide. Greensburg was 1.7 miles wide. The two were perfectly aligned at 9:50 p.m. as the tornado was poised to move north into the town.

About this time, residents were reporting experiencing intense pain in their ears, something like one might feel when diving deep into a swimming pool. That pain was due to the barometric pressure going into a free fall as the tornado bore down on the town.

In Greensburg, the rain and wind ceased. The air felt "freaky," to use one resident's expression. Then, suddenly, houses and buildings began flying apart. Screams filled the air. A highway guardrail went airborne and crashed all the way through a home, through the floor-boards, and into a basement where it fatally impaled a resident. At least one parked automobile went on a high-altitude cross-country flight. A second automobile crashed through the ceiling of one of the local motels, landed on its front bumper, and remained vertical, like a giant avant-garde metal floor lamp. A semitrailer passing through on U.S. 54 reached Greensburg at the same time as the tornado, was blown over, and skidded across the highway. The driver, from Califor-nia, was killed. At 9:54, the winds reached their peak.

At the same moment, on the far north side of the city, the tor-nado reached the Union Pacific tracks. Both train crews gasped as they watched the tornado, illuminated by lightning, cross between them. A second, much smaller tornado spun off from the main storm and crossed the track a moment or two behind its larger brother. The trains, their crews, and the cargo were safe.

The huge tornado started shrinking and went into a death spiral north of town. It turned northwest, west, southwest, and then back southeast in a spiraling counterclockwise path. For a brief moment, it appeared that the tornado might mount a second attack on the now defenseless town. It made another left turn toward the east before reaching Greensburg and then lifted as a second, even larger, tornado touched down just to the east. This second tornado was two miles wide. It headed northeast toward the middle of Kansas.

At 9:58, the Fergusons were eastbound on U.S. 54 and stunned by what they had just seen. As the tornado moved across Greensburg, it was illuminated from the outside by lightning and from the inside by the exploding electrical transformers. Ferguson was on a live call with Merril, telling the KWCH-TV audience that he believed Greensburg had been hit. At 10:00, five minutes after the tornado struck, Merril was reporting that Greensburg had taken "a direct hit." An unknown woman on the fire department radio frequency was shouting, "We need help! We need help!"

In Wichita, Carol Lee, communications director for Midwest LifeTeam, a medical helicopter and aircraft service, heard the TV news about the direct hit and immediately headed to the LifeTeam communications center. A half-hour later, LifeTeam's Dodge City–based helicopter was hovering over Greensburg. As soon as it was safe, other helicopters and aircraft were converging on Greensburg and the other areas hit by the tornado.

Greensburg's emergency services were equally proactive. When it became apparent the town was in danger of a direct hit, emergency officials drove some of the city's emergency vehicles outside of town, out of harm's way. Once the tornado had exited the town, the officials drove back and started rescue efforts.

In marked contrast to the nighttime Udall tornado fifty-two years before, when nearly two hours passed with no assistance, help was on the scene within five minutes of the tornado moving out of Greensburg.

Dick McGowan, a veteran storm chaser from Kansas City, had been following the tornado from the south. Along Highway 183 he encountered one of the tornado's first victims, an injured man, dazed and wandering along the highway. McGowan assisted the man as best he could. Moments later, he received a call from Matt Jacobsen, another chaser, that Greensburg had taken the direct hit. With other chasers holding up downed power lines so McGowan's car could pass, he made it to Greensburg a few minutes later.

Once in town, McGowan's first thought was *Oh, my God.* People were wandering about dazed and in shock. The damage was, in his words, "mind-blowing." And, the smell! For an hour and a half, McGowan and several other chasers went door to door searching for casualties until a sufficient number of trained rescuers could arrive. One of the chasers found a body and brought it to the attention of the authorities.

At 10:00 p.m., as the Fergusons looked in the direction of Greensburg, they saw a solid line of red tail lights. Ferguson realized that traffic was probably stopped at Greensburg, so he used his Storm

Hawk to head north. He wanted to re-intercept the tornado so he could provide additional firsthand reports.

A new hook echo formed on the Dodge City WSR-88 just northeast of Greensburg. This echo was the most classic fishhook shape I've ever seen. And it was lit up in a bright purple color. Since wood, metal, and drywall are more reflective of radar energy than raindrops, what the radar was detecting was elements of what had been the town of Greensburg, thousands of feet above the ground.

As the huge two-mile-wide tornado associated with that hook moved northeast, I called Scott Breit at the office, and we decided we'd give BNSF Railway extra warning time as the tornado approached its track. Amtrak's *Southwest Chief* was scheduled to leave Dodge City eastbound just after midnight. We wanted to ensure that the track was safe and clear before a passenger train moved through the area.

The Fergusons made it north to Kinsley. Lance couldn't get on the Internet, so he ran into a bowling alley to get a look at the radar on the TV stations' continuous coverage and to call KWCH. Kinsley is on U.S. Highway 50, which runs east-west through the town. The radar showed the core of the supercell was passing Highway 50 east of Kinsley, so the couple decided to head east in hopes of picking up the storm.

As they were getting ready to leave Kinsley, the power went out. Ferguson realized the tornado had taken out the electrical transmission line that brought power to Kinsley and points west. He decided to drive slowly, concerned he would encounter debris on their way toward the tornado and, eventually, home to Wichita.

The Fergusons reached the tiny town of Macksville (mistakenly reported to have been hit), which was still intact. Ferguson reported to KWCH-TV and its viewers that Macksville was safe. But two miles east of Macksville, a truck had pulled over. Ferguson pulled over, too. The truck driver pointed into the field where, fifty yards from the highway, there was a ball of metal. At first, Lance did not know what the oddly shaped item was. He noticed the trees near the road were completely shredded. It began to dawn on him that the ball of metal was a car. It was then the paramedics arrived. After speaking with a few of the officials on the scene, the Fergusons moved on and decided to stop chasing for the rest of the night.

From Highway 50, the two drove back south to the town of Pratt. They discussed going to Greensburg to assist with rescue efforts. From a practical standpoint it would have been too dangerous, so they turned east toward Wichita.

As the Fergusons drove toward Wichita, a steady stream of emergency vehicles and school buses (for evacuating the town, if needed) passed them in the opposite direction. Ferguson was struck by how many communities in Kansas, in the middle of the night on a weekend, were rallying to send people and equipment to assist their neighbors in Greensburg.

The series of tornadoes continued all the way into central Kansas. The last of the twisters, number thirteen in the series, lifted southwest of the town of Claflin around 2:10 in the morning, six and a half hours and about 110 miles northeast of where the tornadoes had first started.

Ferguson arrived at KWCH-TV between three and four o'clock in the morning. The station was in its sixth hour of commercial-free continuous storm coverage that would last until 6:00 that Saturday evening. The first thing he asked Merril was, "Do you think we saved any lives tonight?" Merril replied, "I think so. But we won't know until daylight when we see how bad it is."

Ferguson and Merril didn't know that Kiowa County emergency management had put out a call for three refrigerated trucks. Officials were anticipating hundreds of fatalities, and they would need the trucks to store the bodies.

CHAPTER TWENTY-THREE

WHERE THERE'S LIFE, THERE'S HOPE

AMBULANCES, HELICOPTERS, FIRE TRUCKS, HEAVY rescue equipment (for lifting portions of walls, steel girders, etc.), and the media converged on Greensburg from every part of Kansas during the night. The damage was severe; everyone on the scene knew that.

But it took the light of day for the horrible reality to sink in: the town of Greensburg was gone.

The city's beautiful trees were stripped of their leaves, branches, limbs, and bark; instead, eight-foot-high naked white spears populated the town. Ninety-five percent of the town's buildings were destroyed; the other 5 percent damaged. It wasn't just the residents who were in shock; experienced firefighters and cynical newspaper reporters openly wept upon viewing the town's remains.

The Greensburg tornado was one in a thousand: a tornado at the very top of the intensity scale.

Greensburg the morning after the tornado. Courtesy the *Wichita Eagle*.

On May 9, President George W. Bush visited Greensburg. A few hours later, I was on a United Airlines flight from Wichita to Denver. The crew asked permission of air traffic control to deviate to the south of the normal routing. The captain announced we were going to fly over Greensburg and that it would be off to the right side of the plane. Passengers seated left of the aisle got up and looked out the right-side windows as the plane dipped its wing.

A few of my fellow passengers audibly gasped. Several said, "Oh my God." One exclaimed, "there's nothing left." A few looked ill. In May, Kansas' wheat is a beautiful deep green. Lawns are green. Where Greensburg used to be, the dominant color was brown with a slight pinkish hue. Absolutely no green was visible. It resembled a dried bloody scar on the green Kansas countryside.

The story was national news for days. Many speculated that the town might never recover. Greensburg was the county seat of Kiowa County, and the city hall, the police, the fire department, the schools, the grocery store, hospital—absolutely everything—was gone. It wasn't just the buildings; many of the records used by counties and cities to conduct business were gone as well.

Naturally, the media focused on the death toll. Initially it was thought to be seven. And over the following weeks and months, two more people died due to storm injuries, bringing the total deaths in Greensburg to nine. And two others had been killed by other tornadoes from that same supercell later in the night.

One of those later deaths was especially tragic. The unrecognizable automobile viewed by Ferguson near Macksville was the police cruiser of sheriff's officer Tim Buckman. Buckman was out trying to spot the tornado to give an on-the-scene report for the NWS so people in middle Kansas could be warned. Because he didn't have storm-tracking technology in his car, he inadvertently got out ahead of the tornado and into its path. According to the *Wichita Eagle*, his mother had called his cell phone. "Are you all right?" she asked.

"I don't know where I'm at. I can't see anything. It's too late. I'm screwed," were Buckman's last words, and then the phone went dead as the tornado overtook him.

It isn't surprising the media focused on Buckman and the ten others that perished.

But I believe they missed the bigger story: The storm-warning system that has evolved over the past fifty years saved more than 200 lives that night.

How do I know?

The Greensburg and Udall tornadoes were as identical as two tornadoes could possibly be.

Both occurred in the same state—Kansas—so there is similarity in the building codes governing the two towns.

Both occurred well after dark. The two tornadoes approached from the south, rather than the much more common southwest. And because of the southern approach, the tornadoes themselves were obscured by rain and hail.

As the photos in the color insert reveal, the level of damage, right down to the collapsed water towers, was nearly identical. In both cases, 95 percent of the buildings were destroyed, and the other 5 percent were damaged. Both tornadoes were F5 intensity.

Finally, something about the Greensburg radar echoes seemed vaguely familiar. I started comparing the radar tracings of the Udall supercell made at Tinker AFB to the images from the Dodge City radar of the Greensburg storm. The result is in the color insert. The two supercells are virtually identical.

Given that these tornadoes are directly comparable, here's the bottom line: In the Udall tornado, eighty-two people were killed and 260 were injured. The casualty rate in the town was 68 percent. (This doesn't include the twenty people killed in Blackwell from an earlier tornado from the same supercell.)

In Greensburg, with triple the population, nine people were killed and fifty-nine were injured (the tenth and eleventh fatalities were from the next tornado in the series). The casualty rate in Greensburg was less than one-fourteenth that of Udall's.

Put another way, without the warning system, the Greensburg casualty rate would have been the same as Udall's. Given the larger population of Greensburg, that would have meant 243 deaths. Since the actual death toll from all the tornadoes that evening was eleven, 232 people were saved by the warning system. As further evidence, the trained emergency management on the scene expected deaths "in the hundreds," and county commissioner Gene West said, "We had 200 body bags and expected to need them all."

At one point, when the tornado was south of Greensburg, the hook echo folded back into the supercell. Before NEXRAD, that might have been interpreted as a diminishing tornado threat. Because the federal government made the investment in Doppler radar, Umsheid and his NWS colleagues, along with meteorologists at WeatherData and the Wichita TV stations, saw the record rotating winds, knew the tornado threat was increasing, and thus warned accordingly. And

Storm Hawk and other technology provided confirmation of the threat, which added up to a highly credible warning that induced people to take cover.

It was a team effort: WeatherData and the Wichita TV meteorologists. The National Weather Service offices in Dodge City and Wichita. The Storm Prediction Center. The chasers, especially those who helped rescue the injured. Law enforcement and emergency management officials. The quick response by medical teams and rescuers.

In the darkness of a May Friday, the system worked just the way it was supposed to. And we saved 232 lives.

The damage at the north end of Greensburg is remarkable for what it *doesn't* show: a train lying on its side. WeatherData warnings kept the trains out of the path of the tornado. Imagine how much more difficult the rescue and clean-up would have been if the rescuers had to deal with railcars and their contents thrown all over the area.

When I was little, I remember my grandmother saying, "Where there's life, there's hope." Would it be true with Greensburg? Could the town recover and be rebuilt?

* * *

On Sunday, May 4, 2008, the one-year anniversary of the Greensburg tornado, President George W. Bush delivered the commencement address at Greensburg's temporary high school. Under a cloudless sky, Bush told the graduates: "We celebrate your yearlong journey from tragedy to triumph. We celebrate the resurgence of a town that stood tall when its buildings and homes were laid low. We celebrate the power of faith, the love of family, and the bonds of friendship that guided you through the disaster. And finally, we celebrate the resilience of eighteen seniors who grew closer together when the world around them blew apart. When the Class of 2008 walks across the stage today you will send a powerful message to our nation: Greensburg, Kansas, is back—and its best days are ahead."

The president received a warm and appreciative reception from the town.

Knowing the small town would be overrun on the day of the anniversary, photographer Katie Bay and I visited Greensburg the week before to see how the recovery was coming. The town had pledged to rebuild green, with a focus on renewable resources. Signs of progress were everywhere, yet signs of the tornado itself still peppered the town. Brand-new homes stood next to homes that were condemned. Some beautiful homes were rebuilt next to grotesque trees stripped and blanched by the tornado. Signs of progress, but the jury was still out as to whether the town would ultimately be viable.

Kansas is a state that values its pioneer past. While some states have mottoes that sound like tourist marketing slogans, Kansas' motto is *Ad astra per aspera*—"to the stars [greatness] through adversity."

I recalled that motto on the northwest edge of Greensburg, just north of the Union Pacific Railroad tracks, as Katie and I were finishing our visit. We'd ventured into that part of town to photograph a Red Cross trailer in front of a nearly reconstructed home. As we swung the car around to leave, a site caught my eye. We drove about 100 yards to the northeast toward a huge pile of debris. Someone had pulled a tire and rope out of the debris and made a tire swing, and hung it from one of Greensburg's malformed trees.

A good sign: I figure that any town that can make the most of a pile of twisted debris likely has a bright future ahead of it.

EPILOGUE

TWO HUNDRED THIRTY-TWO LIVES SAVED IN A single evening.

Meteorology has come a long way since my dad came running into the house shouting, "Here it comes!"

* * *

Under one of the most spectacular rainbows I have ever observed, people filed into the new gymnasium of the Greensburg High School on October 21, 2009. The Rotary Club of Greensburg had invited me to present *Miracle at Greensburg*, which explains how meteorology saved so many lives that horrible May evening. More than 200 attended. The audience audibly gasped when they saw the graphic of the Greensburg radar echoes superimposed over the Udall radar tracings. When I introduced meteorologist Mike Umshied of the Dodge City National Weather Service (who issued the tornado warning for

Greensburg) and WeatherData meteorologists Scott Breit and Trevor LaVoie (who issued the warning that allowed Union Pacific to keep its trains safely out of the area), the crowd gave them a standing ovation and a long, appreciative round of applause.

After my formal presentation ended, I spent nearly an hour chatting with various members of the community. They fully realize how much worse things would have been without the warnings.

A look around Greensburg today reveals a sparkling arts center (recently pictured in the *New York Times*), new stores and businesses, and a town dedicated to rebuilding "green." A new wind farm was dedicated two days after my visit. The town is also using solar, geothermal, and other green energy sources. At least one business is relocating from Texas to Greensburg. The town is being reborn. Hope has become reality.

* * *

I am pleased to report that meteorology is not waiting another forty years to design the next generation of radars. With the NEXRADs about a dozen years old, the National Severe Storms Lab has developed dual-polarization radar that will be a major near-term upgrade and should improve our ability to warn of flash floods and winter storms in much the same way as the original NEXRADs have contributed to tornadoes and severe thunderstorm warnings.

In a few years, we will likely see phased array radar replace the NEXRADs, allowing us to get a three-dimensional view of storms and further increasing the timeliness and accuracy of storm warnings and aviation-weather advisories.

A new generation of weather satellites to be launched in the second decade of the twenty-first century should improve our ability to measure the atmosphere. The improved measurements, combined with improved computer models, will allow us to increase forecasting and warning precision.

The Vortex II scientific experiment in 2009 collected the most comprehensive data about a single tornado in history. We are hopeful that data will help us better understand the inner workings of tornadoes which should lead to even more accurate forecasts and warnings.

While we'll never reach forecasting perfection, I am confident that meteorology will be up to the challenges that future storms may pose.

For me, preparedness is the issue of greater concern. We didn't apply the lessons of Andrew to Katrina. Will we learn from the Katrina fiasco?

Are we really prepared to conduct an evacuation of New York City when the inevitable Category 4 storm threatens? Do we have the infrastructure to get all of those people out in a timely and orderly fashion? Given the importance of Wall Street to our economy, will financial markets be able to weather weeks or months of disruption such a storm would cause? Similar questions could be asked about Houston and the energy industry, as well as other major cities in coastal areas.

* * *

In the meantime, I continue to enjoy a wonderful career in meteorology and have the privilege of working with some supremely talented men and women. The closest I have ever come to fainting occurred when I was told that I was being awarded the American Meteorological Society's Award for Outstanding Contribution to Applied Meteorology for my work in the field of storm warnings and newspaper weather. It was even more amazing when WeatherData received the Award for Outstanding Service to Meteorology by a Corporation in 2000. No other organization has those two awards in its trophy case. Our parent company, AccuWeather, received the Corporation award for 2010, further testimony to the outstanding work we do.

Many of my colleagues have not received the recognition I have enjoyed but are no less deserving. If you appreciate the work your favorite TV meteorologist does, drop him or her a note or, better yet, drop a note to the station's general manager. If your local NWS office issues a particularly good storm warning that is useful to you, drop those folks a note as well.

In March of 2006, I sold the assets of WeatherData to AccuWeather, Inc., so we would have a larger base upon which to build our business. That has worked out well for both parties, and I have enjoyed new business challenges as I have worked with my new colleagues to build a bigger WeatherData and apply my talents to our parent company.

Best of all, I have been blessed with an amazingly supportive wife, three terrific children, and a wonderful daughter-in-law. I could not be more proud of them all.

And that brings me to one last story to share with you.

* * *

May 19, 2007, fifteen days after Greensburg. It was a bright, sunny, glorious morning. Our daughter, Tiffany, was graduating from the University of Kansas School of Journalism in Lawrence. It was a touching ceremony for the proud parents, brothers, sister-in-law, and grandparents. Our youngest, our baby girl, was fully grown up and graduating from college. Or, when I'm feeling ornery, I say, "they are all off the payroll."

But there was something else I had to attend to after the graduation festivities. I drove from Lawrence to Ruskin Heights for the fiftieth anniversary commemoration of that terrible night. It had been years since I had been in the area.

The commemoration was amazingly well done. People shared their stories of survival, heartbreak, and heroism from that evening. Joe Audsley, looking frail, recounted his incredible work that had saved so many lives.

The emotionalism deepened as the afternoon unfolded. The Lees Summit High School Drama Club presented a play about that night, which could have been hokey but was, instead, deeply moving.

Then the school board presented Ruskin High School diplomas to the children who'd been killed before they could graduate. I met Bobbi Davis, the woman who as a girl had been in the car thrown against the top of the water tower. She accepted a diploma on her sister's behalf.

We went outside to rededicate a memorial to that day. A bell was rung forty-four times: once for each of the people who'd lost their lives. As we walked back inside, I couldn't help but notice a tornado siren on the school grounds. Should another tornado approach the school, the siren will ensure that everyone inside will know it is coming.

As I drove away, there was one more thing I needed to do: I drove back to the old neighborhood, checked out the old house, and for the first time since I was six years old, I walked through the Dead End. At the age of fifty-five, the first thing that hit me was, "It's so small!" When I was five and six years old, I thought it was huge.

As I walked toward the railroad embankment and crossed to the south side of the little creek, I felt it: a small, growing vibration through my shoes. Instantly, I was transported back fifty years—a train was coming!

Although the Kansas City Southern Railroad's elevated right-of-way is now overgrown with trees and foliage and does not have the clear view of the track I enjoyed in my childhood, I could see a Kansas City Southern locomotive heading my way. It was one of three that had been painted in Southern Belle heritage colors. Behind it, helping pull the train, was a Southern Pacific locomotive—the first railroad WeatherData ever served.

In some ways, my world had come full circle in a single day.

* * *

Meteorology took its first tentative steps at tornado and hurricane warnings a little more than fifty years ago. From that inauspicious beginning involving spare World War II leftovers, we have developed an effective and highly cost-effective system that saves lives and dollars, nearly every week.

Good science does not have to take billions of dollars. It requires dedication, outside-the-box thinking, and a willingness to go where the data and experimental results might take you.

So, here's to the first fifty years of storm warnings. May the next fifty be just as rewarding.

ABOUT THE AUTHOR

MIKE SMITH knew he would be a meteorologist at the age of five when a major tornado occurred near his Kansas City home. Fifty years later, Smith has become one of America's most innovative and honored meteorologists.

Considered a pioneer in the meteorology world, Smith's development of the color radar literally "colored the weather," and he was one of the nation's first storm chasers. After receiving his meteorology degree from the top-ranked university for severe storm research, the University of Oklahoma, he worked as a television meteorologist in major markets, including St. Louis, Oklahoma City, and Wichita. During his time in television, Smith became the first person ever to do a live telecast of a tornado, demonstrating the ruthlessness Mother Nature can bring.

In 1981, Smith founded WeatherData Services, Inc., a company credited with saving countless lives and preventing hundreds of millions of dollars in property losses. WeatherData Services was the first company to provide customized severe weather warnings directly to businesses such as railroads, airlines, and other industries with the potential to be severely impacted by weather damages. He has received eighteen U.S. and foreign patents in the fields of weather science, emergency management, and search and rescue, which have been used to foster advancements in the field of meteorology.

As a Fellow of the American Meteorological Society (AMS), Smith received the prestigious award for Outstanding Contribution to the Advance of Applied Meteorology for his work in severe weather warnings and newspaper weather displays. WeatherData and its parent company, AccuWeather, have both been recipients of the AMS's Award for Outstanding Services to Meteorology by a Corporation. Smith is the only individual in applied meteorology to have received this level of recognition and honor.

In addition to his work at WeatherData, he is a frequent speaker and author of both popular and technical weather-related articles. He has appeared on the Discovery Channel, the History Channel, *Fox Business News*, *Today*, *NBC Nightly News*, *CBS Evening News*, and numerous other media outlets and is the author of the weather blog meteorologicalmusings.

To inquire about booking Mike Smith as a keynote speaker for your event, please go to www.mikesmithenterprises.com and contact Kim Dugger, Director of Marketing, Mike Smith Enterprises, LLC.